T0327856

REDEFINING DIVERSITY AND DYNAMICS OF NATURAL RESOURCES MANAGEMENT IN ASIA

VOLUME 3

Natural Resource Dynamics and Social Ecological System in Central Vietnam: Development, Resource Changes and Conservation Issues
Tran Nam Thang, Ngo Tri Dung, David Hulse, Shubhechchha Sharma and Ganesh P. Shivakoti; editors

Volume 3 is dedicated to Nobel Laureate Elinor Ostrom who is the source of inspiration in drafting these volumes and all chapter authors of these volumes have benefited from her theoretical framework.

Book Title: Re-defining Diversity and Dynamism of Natural Resource Management in Asia

Book Editors: Ganesh P. Shivakoti, Shubhechchha Sharma and Raza Ullah

Other volumes published:

1) Sustainable Natural Resources Management in Dynamic Asia Volume 1
   Ganesh P. Shivakoti, Ujjwal Pradhan and Helmi; editors
2) Upland Natural Resources and Social Ecological Systems in Northern Vietnam Volume 2
   Mai Van Thanh, Tran Duc Vien, Stephen J. Leisz and Ganesh P. Shivakoti; editors
3) Reciprocal Relationship between Governance of Natural Resources and Socio-Ecological Systems Dynamics in West Sumatra, Indonesia Volume 4

Rudi Febriamansyah, Yonariza, Raza Ullah and Ganesh P. Shivakoti; editors

REDEFINING DIVERSITY AND DYNAMICS OF NATURAL RESOURCES MANAGEMENT IN ASIA

Natural Resource Dynamics and Social Ecological Systems in Central Vietnam: Development, Resource Changes and Conservation Issues

VOLUME 3

EDITED BY

TRAN NAM THANG, NGO TRI DUNG, DAVID HULSE,

SHUBHECHCHHA SHARMA AND GANESH P. SHIVAKOTI

ELSEVIER AMSTERDAM • BOSTON • HEIDELBERG • LONDON • NEW YORK • OXFORD
PARIS • SAN DIEGO • SAN FRANCISCO • SINGAPORE • SYDNEY • TOKYO

Elsevier
Radarweg 29, PO Box 211, 1000 AE Amsterdam, Netherlands
The Boulevard, Langford Lane, Kidlington, Oxford OX5 1GB, United Kingdom
50 Hampshire Street, 5th Floor, Cambridge, MA 02139, United States

Notices
Knowledge and best practice in this field are constantly changing. As new research and experience broaden our
understanding, changes in research methods, professional practices, or medical treatment may become necessary.

Practitioners and researchers must always rely on their own experience and knowledge in evaluating and using
any information, methods, compounds, or experiments described herein. In using such information or methods
they should be mindful of their own safety and the safety of others, including parties for whom they have a
professional responsibility.

To the fullest extent of the law, neither the Publisher nor the authors, contributors, or editors, assume any liability
for any injury and/or damage to persons or property as a matter of products liability, negligence or otherwise, or
from any use or operation of any methods, products, instructions, or ideas contained in the material herein.

Library of Congress Cataloging-in-Publication Data
A catalog record for this book is available from the Library of Congress

British Library Cataloguing-in-Publication Data
A catalogue record for this book is available from the British Library

ISBN: 978-0-12-805452-9

For information on all Elsevier publications
visit our website at https://www.elsevier.com/

  Working together
to grow libraries in
Book Aid International developing countries

www.elsevier.com • www.bookaid.org

Publisher: Candice G. Janco
Acquisition Editor: Laura S Kelleher
Editorial Project Manager: Emily Thomson
Production Project Manager: Mohanapriyan Rajendran
Cover Designer: Matthew Limbert

Typeset by SPi Global, India

Contents

VI

CONCLUDING SECTION

Contributors

H.V. Chuong University of Agriculture and Forestry, Hue University, Hue City, Vietnam

R. Cochard Institute of Integrative Biology, Swiss Federal Institute of Technology, Zurich, Switzerland; Asian Institute of Technology, Klong Luang, Pathumthani, Thailand

N.T. Duc Hue University of Agriculture and Forestry, Hue City, Vietnam

N.T. Dung Hue University of Agriculture and Forestry, Hue, Vietnam

D.T. Duong Hue University of Agriculture and Forestry, Hue, Vietnam

H.D. Ha Hue University of Agriculture and Forestry, Hue, Vietnam

N.X. Hong Hue University of Science, Hue, Vietnam

D. Hulse Ford Foundation, Jakarta, Indonesia

M.Q. Huy Hue Forest Protection Department, Hue, Vietnam

N.T.T. Lien Hue University of Sciences, Hue, Vietnam

T.D. Ngo Hue University of Agriculture and Forestry, Hue, Vietnam

P.T. Nhung Hue University of Agriculture and Forestry, Hue, Vietnam

T.T. Phuong University of Agriculture and Forestry, Hue University, Hue City, Vietnam

H.T.A. Phuong Hue University of Sciences (HUS), Hue, Vietnam

H.T.H. Que Hue University of Agriculture and Forestry, Hue, Vietnam

S. Sharma WWF Nepal, Kathmandu, Nepal

G. Shivakoti The University of Tokyo, Tokyo, Japan; Asian Institute of Technology, Bangkok, Thailand

R.P. Shrestha Asian Institute of Technology, Pathumthani, Thailand

T.N. Thang Hue University of Agriculture and Forestry, Hue, Vietnam

Thiha The Treedom Group, Bangkok, Thailand

N.D.A. Tuan Forest Protection Department of Thua Thien Hue, Vietnam

L. Van An Hue University of Agriculture and Forestry, Hue City, Vietnam

L. Van Lan Hue University of Agriculture and Forestry, Hue, Vietnam

E.L. Webb National University of Singapore, Singapore

V.T. Yen Hue University of Agriculture and Forestry, Hue, Vietnam

Words From Book Editors

CONTEXT

Elinor Ostrom received the Nobel Prize in Economics for showing how the "commons" is vital to the livelihoods of many throughout the world. Her work examined the rhetoric of the "tragedy of the commons," which has been used as the underlying foundation in privatizing property and centralizing its management as a way to protect finite resources from depletion. She worked, along with others, to overturn the "conventional wisdom" of the tragedy of the commons by validating the means and ways that local resources can be effectively managed through common property regimes instead of through the central government or privatization. Ostrom identified eight design principles relating to how common pool resources can be governed sustainably and equitably in a community. Similarly, the Institutional Analysis and Development (IAD) framework summarizes the ways in which institutions function and adjust over time. The framework is a "multi-level conceptual map," which describes a specific hierarchical section of interactions made in a system. The framework seeks to identify and explain interactions between actors and action situations.

As a political scientist, Ostrom has been a source of inspiration for many researchers and social scientists, including this four volumes book. Her theories and approach serve as the foundation for many of the chapters within these volumes. Following in her footsteps, the books is based on information collected during fieldwork that utilized quantitative as well as qualitative data, and on comparative case studies, which were then analyzed to gain an understanding of the situation, rather than starting from a formulated assumption of reality. The case studies in these volumes highlight the issues linked to the management of the environment and natural resources, and seek to bring about an understanding of the mechanisms used in managing the natural resource base in the regions, and how different stakeholders interact with each other in managing these natural resources. The details of the books are as follows:

Volume title		Editors
"Re-defining Diversity and Dynamism of Natural Resources Management in Asia"		Ganesh P. Shivakoti, Shubhechchha Sharma, and Raza Ullah
Volume I	Sustainable Natural Resources Management in Dynamic Asia	Ganesh P. Shivakoti, Ujjwal Pradhan, and Helmi
Volume II	Upland Natural Resources and Social Ecological Systems in Northern Vietnam	Mai Van Thanh, Tran Duc Vien, Stephen J. Leisz, and Ganesh P. Shivakoti
Volume III	Natural Resource Dynamics and Social Ecological Systems in Central Vietnam: Development, Resource Changes and Conservation Issue	Tran Nam Thang, Ngo Tri Dung, David Hulse, Shubhechchha Sharma, and Ganesh P. Shivakoti

Continued

Volume title		Editors
Volume IV	Reciprocal Relationship between Governance of Natural Resources and Socio-Ecological Systems Dynamics in West Sumatra Indonesia	Rudi Febriamansyah, Yonariza, Raza Ullah, and Ganesh P. Shivakoti

These volumes are made possible through the collaboration of diverse stakeholders. The intellectual support provided by Elinor Ostrom and other colleagues through the Ostrom Workshop in Political Theory and Policy Analysis at the Indiana University over the last two and half decades has provided a solid foundation for drafting the book. The colleagues at the Asian Institute of Technology (AIT) have been actively collaborating with the Workshop since the creation of the Nepal Irrigation, Institutions and Systems (NIIS) database; and the later Asian Irrigation, Institutions and Systems (AIIS) database (Ostrom, Benjamin and Shivakoti, 1992; Shivakoti and Ostrom, 2002; Shivakoti et al., 2005; Ostrom, Lam, Pradhan and Shivakoti, 2011). The International Forest Resources and Institutions (IFRI) network carried out research to support policy makers and practitioners in designing evidence based natural resource polices based on the IAD framework at Indiana University, which was further mainstreamed by the University of Michigan. In order to support this, the Ford Foundation (Vietnam, India, and Indonesia) provided grants for capacity building and concerted knowledge sharing mechanisms in integrated natural resources management (INRM) at Indonesia's Andalas University in West Sumatra, Vietnam's National University of Agriculture (VNUA) in Hanoi, and the Hue University of Agriculture and Forestry (HUAF) in Hue, as well as at the AIT for collaboration in curric-

ulum development and in building capacity through mutual learning in the form of masters and PhD fellowships (Webb and Shivakoti, 2008). Earlier, the MacArthur Foundation explored ways to support natural resource dependent communities through the long term monitoring of biodiversity, the domestication of valuable plant species, and by embarking on long-term training programs to aid communities in managing natural resources.

VOLUME 1

This volume raises issues related to the dependence of local communities on natural resources for their livelihood; their rights, access, and control over natural resources; the current practices being adopted in managing natural resources and socio-ecological systems; and new forms of natural resource governance, including the implementation methodology of REDD+ in three countries in Asia. This volume also links regional issues with those at the local level, and contributes to the process of application of various multimethod and modeling techniques and approaches, which is identified in the current volume in order to build problem solving mechanisms for the management of natural resources at the local level. Earlier, the Ford Foundation Delhi office supported a workshop on Asian Irrigation in Transition, and its subsequent publication (Shivakoti et al., 2005) was followed by Ford Foundation Jakarta office's long term support for expanding the knowledge on integrated natural resources management, as mediated by institutions in the dynamic social ecological systems.

VOLUME 2

From the early 1990s to the present, the Center for Agricultural Research and

Ecological Studies (CARES) of VNUA and the School of Environment, Resources and Development (SERD) of AIT have collaborated in studying and understanding the participatory process that has occurred during the transition from traditional swidden farming to other farming systems promoted as ecologically sustainable, livelihood adaptations by local communities in the northern Vietnamese terrain, with a special note made to the newly emerging context of climate change. This collaborative effort, which is aimed at reconciling the standard concepts of development with conservation, has focused on the small microwatersheds within the larger Red River delta basin. Support for this effort has been provided by the Ford Foundation and the MacArthur Foundation, in close coordination with CARES and VNUA, with the guidance from the Ministry of Agriculture and Rural Development (MARD) and the Ministry of Natural Resources and Environment (MONRE) at the national, regional, and community level. Notable research documentation in this volume includes issues such as local-level land cover and land use transitions, conservation and development related agro-forestry policy outcomes at the local level, and alternative livelihood adaptation and management strategies in the context of climate change. A majority of these studies have examined the outcomes of conservation and development policies on rural communities, which have participated in their implementation through collaborative governance and participatory management in partnership with participatory community institutions. The editors and authors feel that the findings of these rich field-based studies will not only be of interest and use to national policymakers and practitioners and the faculty and students of academic institutions, but can also be equally applicable to guiding conservation and development issues for those schol-ars interested in understanding a developing country's social ecological systems, and its context-specific adaptation strategies.

VOLUME 3

From the early 2000 to the present, Hue University of Agriculture and Forestry (HUAF) and the School of Environment, Resources and Development (SERD) of AIT supported by MacArthur Foundation and Ford Foundation Jakarta office have collaborated in studying and understanding the participatory process of Social Ecological Systems Dynamics that has occurred during the opening up of Central Highland for infrastructure development. This collaborative effort, which is aimed at reconciling the standard concepts of development with conservation, has focused on the balance between conservation and development in the buffer zone areas as mediated by public resource management institutions such as Ministry of Agriculture and Rural Development (MARD), Ministry of Natural Resources and Environment (MONRE) including National Parks located in the region. Notable research documentation in this volume includes on issues such as local level conservation and development related policy outcomes at the local level, alternative livelihood adaptation and management strategies in the context of climate change. A majority of these studies have examined the outcomes of conservation and development policies on the rural communities which have participated in their implementation through collaborative governance and participatory management in partnership with participatory community institutions.

VOLUME 4

The issues discussed above are pronounced more in Indonesia among the Asian countries and the Western Sumatra is such typical example mainly due to earlier logging concessionaries, recent expansion of State and private plantation of para-rubber and oil palm plantation. These new frontiers have created confrontations among the local community deriving their livelihoods based on inland and coastal natural resources and the outsiders starting mega projects based on local resources be it the plantations or the massive coastal aqua cultural development. To document these dynamic processes Ford Foundation Country Office in Jakarta funded collaborative project between Andalas University and Asian Institute of Technology (AIT) on Capacity building in Integrated Natural Resources Management. The main objective of the project was Andalas faculty participate in understanding theories and diverse policy arenas for understanding and managing common pool resources (CPRs) which have collective action problem and dilemma through masters and doctoral field research on a collaborative mode (AIT, Indiana University and Andalas). This laid foundation for joint graduate program in Integrated Natural Resources Management (INRM). Major activities of the Ford Foundation initiatives involved the faculty from Andalas not only complete their degrees at AIT but also participated in several collaborative training.

1 BACKGROUND

Throughout Asia, degradation of natural resources is happening at a higher rate, and is a primary environmental concern. Recent tragedies associated with climate change have left a clear footprint on them, from deforestation, land degradation, and changing hydrological and precipitation patterns. A significant proportion of land use conversion is undertaken through rural activities, where

resource degradation and deforestation is often the result of overexploitation by users who make resource-use decisions based on a complex matrix of options, and potential outcomes.

South and Southeast Asia are among the most dynamic regions in the world. The fundamental political and socioeconomic setting has been altered following decades of political, financial, and economic turmoil in the region. The economic growth, infrastructure development, and industrialization are having concurrent impacts on natural resources in the form of resource degradation, and the result is often social turmoil at different scales. The natural resource base is being degraded at the cost of producing economic output. Some of these impacts have been offset by enhancing natural resource use efficiency, and through appropriate technology extension. However, the net end results are prominent in terms of increasing resource depletion and social unrest. Furthermore, climate change impacts call for further adaptation and mitigation measures in order to address the consequences of erratic precipitation and temperature fluctuations, salt intrusions, and sea level increases which ultimately affect the livelihood of natural resource dependent communities.

Governments, Non-governmental organizations (NGOs), and academics have been searching for appropriate policy recommendations that will mitigate the trend of natural resource degradation. By promoting effective policy and building the capacity of key stakeholders, it is envisioned that sustainable development can be promoted from both the top-down and bottom-up perspectives. Capacity building in the field of natural resource management, and poverty alleviation is, then, an urgent need; and several policy alternatives have been suggested (Inoue and Shivakoti, 2015; Inoue and Isozaki, 2003; Webb and Shivakoti, 2008).

The importance of informed policy guidance in sustainable governance and the management of common pool resources (CPRs), in general, have been recognized due to the conflicting and competing demand for use of these resources in the changing economic context in Asia (Balooni and Inoue, 2007; Nath, Inoue and Chakma, 2005; Pulhin, Inoue and Enters, 2007; Shivakoti and Ostrom, 2008; Viswanathan and Shivakoti, 2008). This is because these resources are unique in respect to their context. The management of these resources are by the public, often by local people, in a partnership between the state and the local community; but on a day-to-day basis, the benefits are at the individual and private level. In the larger environmental context, however, the benefits and costs have global implications. There are several modes of governance and management arrangement possible for these resources in a private-public partnership. Several issues related to governance and management need to be addressed, which can directly feed into the ongoing policy efforts of decentralization and poverty reduction measures in South and South East Asia.

While there has been a large number of studies, and many management prescriptions made, for the management of natural resources, either from the national development point-of-view or from the local-level community perspectives, there are few studies which point toward the interrelationship among other resources and CPRs, as mediated by institutional arrangement, and that have implications for the management of CPRs in an integrated manner, vis-a-vis poverty reduction. In our previous research, we have identified several anomalies and tried to explain these in terms of better management regimes for the CPRs of several Asian countries (Dorji, Webb and Shivakoti, 2006; Gautam, Shivakoti and Webb, 2004; Kitjewachakul, Shivakoti and Webb, 2004;

Mahdi, Shivakoti and Schmidt-Vogt, 2009; Shivakoti et al., 1997; Dung and Webb, 2008; Yonariza and Shivakoti, 2008). However, there are still several issues, such as the failure to comprehend and conceptualize social and ecological systems as coupled systems that adapt, self-organize, and are coevolutionary. The information obtained through these studies tends to be fragmented and scattered, leading to incomplete decision making, as they do not reflect the entire scenario. The shared vision of the diverse complexities, that are the reality of natural resource management, needs to be fed into the governance and management arrangements in order to create appropriate management guidelines for the integrated management of natural resources, and CPR as a whole.

Specifically, the following issues are of interest:

a. How can economic growth be encouraged while holding natural resources intact?

b. How has the decentralization of natural management rights affected the resource conditions, and how has it addressed concerns of the necessity to incorporate gender concerns and social inclusion in the process?

c. How can the sustainability efforts to improve the productive capacity of CPR systems be assessed in the context of the current debate on the effects of climate change, and the implementation of new programs such as Payment for Ecosystem Services (PES) and REDD+?

d. How can multiple methods of information gathering and analysis (eg use of various qualitative and quantitative social science methods in conjunction with methods from the biological sciences, and time series remote sensing data collection methods) on CPRs be integrated into national natural resource policy guidelines, and the results be used by local managers and users of CPRs, government agencies, and scholars?

e. What are the effective polycentric policy approaches for governance and management of CPRs, which are environmentally sustainable and gender balanced?

2 OBJECTIVES OF THESE VOLUMES

At each level of society, there are stakeholders, both at the public and private level, who are primarily concerned with efforts of management enhancement and policy arrangements. Current theoretical research indicates that this is the case whether it is deforestation, resource degradation, the conservation of biodiversity hotspots, or climate change adaptation. The real struggles of these local-level actors directly affect the management of CPR, as well as the hundreds of people who are dependent upon them for a living. This book is about those decisions as the managers of natural resources. Basically, the authors of these chapters explore outcomes after decentralization and economic reforms, respectively. The volumes of this book scrutinize the variations of management practices with, and between, communities, local administration, and the CPR. Economic growth is every country's desire, but in the context of South and South East Asia, much of the economic growth is enabled by the over use of the natural resource base. The conundrum is that these countries need economic growth to advance, but the models of economic growth that are advanced, negatively affect the environment, which the country, depends upon. Examples of this are seen in such varied contexts as the construction of highways through protected areas, the construction of massive hydropower dams, and the conversion of traditional agricultural fields into rubber and oil palm plantations.

The research also shows that the different levels of communities, administration, and people are sometimes highly interactive and overlapping, for that reason, it is necessary to undertake coordinated activities that lead to information capture and capacity building at the national, district, and local levels. Thus the impacts of earlier intervention efforts (various policies in general and decentralization in particular) for effective outcomes have been limited, due to the unwillingness of higher administrative officials to give up their authority, the lack of trust and confidence of officials in the ability of local communities in managing CPR, local elites capturing the benefits of decentralization in their favor, and high occurrences of conflicts among multiple stakeholders at the local level (IGES, 2007).

In the areas of natural resource management particular to wildlife ecology monitoring and climate change adaptation, the merging of traditional knowledge with science is likely to result in better management results. Within many societies, daily practices and ways of life are constantly changing and adapting to new situations and realities. Information passed through these societies, while not precise and usually of a qualitative nature, is valued for the reason that it is derived from experience over time. Scientific studies can backstop local knowledge, and augment it through the application of rigorous scientific method derived knowledge, examining the best practices in various natural resource management systems over spatial and temporal scales. The amalgamation of scientific studies and local knowledge, which is trusted by locals, may lead to powerful new policies directed toward nature conservation and livelihood improvements.

Ethnic minorities, living in the vicinity to giant infrastructure projects, have unequal access, and control over, resources compared to other more powerful groups. Subsistence agriculture, fishery, swiddening, and a few off-farm options are the livelihood activities for these individuals. But unfortunately, these livelihood options are in areas that will be hit the most by changing climatic scenarios, and these people are the least equipped to cope; a situation that further aggravates the possibility of diversifying their livelihood options. Increasing tree coverage can help to mitigate climate change through the sequestering of carbon in trees. Sustainably planting trees requires technical, social, and political dimensions that are mainly possible through the decentralization of power to local communities to prevent issues of deforestation and degradation. The role of traditional institutions hence becomes crucial to reviving social learning, risk sharing, diversifying options, formulating adaptive plans and their effective implementation, fostering stress tolerance, and capacity building against climate change effects.

Though, the role of institutions in managing common pool resources has been explained in literature, it is also worth noting that institutions play significant roles in climate change adaptation. A study conducted by Gabunda and Barker (1995) and Nyangena (2004) observed that household affiliations in social networks were highly correlated with embracing soil erosion retaining technologies. Likewise, Jagger and Pender (2006) assumed that individuals involved in natural resource management focused programs were likely to implement land management expertise, regardless of their direct involvement in particular organizations. Friis-Hansen (2005) partially verifies that there is a positive relationship among participation in a farmer's institution and the adoption of smart agriculture technology. Dorward et al. (2009) correspondingly notes that institutions are vital in shaping the capability of local agrarians to respond to challenges and opportunities. This study has also shown that institutions are the primary attribute in fostering individuals and households to diversify livelihoods in order to adapt to a changing climate. In the context of REDD+, a system is required that can transcend national boundaries, interconnect different governance levels, and allow both traditional and modern policy actors to cooperate. Such a system emphasizes the integration of both formal and informal rule making mechanisms and actor linkages in every governance stage, which steer toward adapting to and mitigating the effects of local and global environmental change (Corbera and Schroeder, 2010).

Based on the above noted discussions, the volumes in this book bring these issues forward for a global audience and policy makers. Though earlier studies show that the relationship between scientific study and outcomes in decision making are usually complex; we hope that the studies examined and discussed here can have some degree of impact on academics, practitioners, and managers.

G. Shivakoti, S. Sharma, and R. Ullah

References

Balooni, K.B., Inoue, M., 2007. Decentralized forest management in South and Southeast Asia. J. Forest. 2007, 414–420.

Corbera, E., Schroeder, H., 2010. Governing and Implementing REDD+. Environ. Sci. Pol. http://dx.doi.org/10.1016/j.envsci.2010.11.002.

Dorji, L., Webb, E., Shivakoti, G.P., 2006. Forest property rights under nationalized forest management in Bhutan. Environ. Conservat. 33 (2), 141–147.

Dorward, A., Kirsten, J., Omamo, S., Poulton, C., Vink, N., 2009. Institutions and the agricultural development challenge in Africa. In: Kirsten, J.F., Dorward, A.R., Poulton, C., Vink, N. (Eds.), Institutional Economics Perspectives on African Agricultural Development. IFPRI, Washington DC.

Dung, N.T., Webb, E., 2008. Incentives of the forest land allocation process: Implications for forest management in Nam Dong District, Central Vietnam. In: Webb, E., Shivakoti, G.P. (Eds.), Decentralization, Forests and Rural Communities: Policy outcome in South and South East Asia. SAGE Publications, New Delhi, pp. 269–291.

Friis-Hansen, E., 2005. Agricultural development among poor farmers in Soroti district, Uganda: Impact Assessment of agricultural technology, farmer empowerment and changes in opportunity structures. Paper presented at Impact Assessment Workshop at CYMMYT, Mexico, 19–21. October. http://citeseerx.ist.psu.edu/viewdoc/download?doi=10.1.1.464.8651&rep=rep1&type=pdf.

Gautam, A., Shivakoti, G.P., Webb, E.L., 2004. A review of forest policies, institutions, and the resource condition in Nepal. Int. Forest. Rev. 6 (2), 136–148.

Gabunda, F., Barker, R., 1995. Adoption of hedgerow technology in Matalom, Leyte Philipines. Mimeo. In: Bluffstone, R., Khlin, G. (Eds.), 2011. Agricultural Investment and Productivity: Building Sustainability in East Africa. RFF Press, Washington, DC/London.

IGES, 2007. Decentralization and State-sponsored Community Forestry in Asia. Institute for Global Environmental Studies, Kanagawa.

Inoue, M., Isozaki, H., 2003. People and Forest-policy and Local Reality in Southeast Asia, the Russian Far East and Japan. Kluwer Academic Publishers, Netherlands.

Inoue, M., Shivakoti, G.P. (Eds.), 2015. Multi-level Forest Governance in Asia: Concepts, Challenges and the Way Forward. Sage Publications, New Delhi/California/London/Singapore.

Jagger, P., Pender, J., 2006. Impacts of Programs and Organizations on the Adoption of Sustainable Land Management Technologies in Uganda. IFPRI, Washington, DC.

Kijtewachakul, N., Shivakoti, G.P., Webb, E., 2004. Forest health, collective behaviors and management. Environ. Manage. 33 (5), 620–636.

Mahdi, Shivakoti, G.P., Schmidt-Vogt, D., 2009. Livelihood change and livelihood sustainability in the uplands of Lembang Subwatershed, West Sumatra, Indonesia, in a changing natural resource management context. Environ. Manage. 43, 84–99.

Nath, T.K., Inoue, M., Chakma, S., 2005. Prevailing shifting cultivation in the Chittagong Hill Tracts, Bangladesh: some thoughts on rural livelihood and policy issues. Int. For. Rev. 7 (5), 327–328.

Nyangena, W., 2004. The effect of social capital on technology adoption: empirical evidence from Kenya. Paper presented at 13th Annual Conference of the European Association of Environmental and Resource Economics, Budapest.

Ostrom, E., Benjamin, P., Shivakoti, G.P., 1992. Institutions, Incentives, and Irrigation in Nepal: June 1992. (Monograph) Workshop in Political Theory and Policy Analysis, Indiana University, Bloomington, Indiana, USA.

Ostrom, E., Lam, W.F., Pradhan, P., Shivakoti, G.P., 2011. Improving Irrigation Performance in Asia: Innovative Intervention in Nepal. Edward Elgar Publishers, Cheltenham, UK.

Pulhin, J.M., Inoue, M., Enters, T., 2007. Three decades of community-based forest management in the Philippines: emerging lessons for sustainable and equitable forest management. Int. For. Rev. 9 (4), 865–883.

Shivakoti, G., Ostrom, E., 2008. Facilitating decentralized policies for sustainable governance and management of forest resources in Asia. In: Webb, E., Shivakoti, G.P. (Eds.), Decentralization, Forests and Rural Communities: Policy Outcomes in South and Southeast Asia. Sage Publications, New Delhi/Thousand Oaks/London/Singapore, pp. 292–310.

Shivakoti, G.P., Ostrom, E. (Eds.), 2002. Improving Irrigation Governance and Management in Nepal. Institute of Contemporary Studies (ICS) Press, California, Oakland.

Shivakoti, G.P., Vermillion, D., Lam, W.F., Ostrom, E., Pradhan, U., Yoder, R., 2005. Asian Irrigation in Transition-Responding to Challenges. Sage Publications, New Delhi/Thousand Oaks/London.

Shivakoti, G., Varughese, G., Ostrom, E., Shukla, A., Thapa, G., 1997. People and participation in sustainable development: understanding the dynamics of natural resource system. In: Proceedings of an International Conference held at Institute of Agriculture and Animal Science, Rampur, Chitwan, Nepal. 17–21 March, 1996. Bloomington, Indiana and Rampur, Chitwan.

Viswanathan, P.K., Shivakoti, G.P., 2008. Adoption of rubber integrated farm livelihood systems: contrasting empirical evidences from Indian context. J. For. Res. 13 (1), 1–14.

Webb, E., Shivakoti, G.P. (Eds.), 2008. Decentralization, Forests and Rural Communities: Policy Outcomes in South and Southeast Asia. Sage Publications, New Delhi/Thousand Oaks/London/Singapore.

Yonariza, Shivakoti, G.P., 2008. Decentralization and co-management of protected areas in Indonesia. J. Legal Plur. 57, 141–165.

Foreword

For more than two decades, Vietnam has been one of the most dynamic and rapidly developing countries in Southeast Asia. The emergence of a market-based economy resulting from the introduction of economic reforms (đổi mới) in the mid-1980s resulted in rapid economic gains and poverty reduction. According to the World Bank, extreme poverty has been reduced to only about 3% of the population from about 50% in the 1990s. Similarly, the per-capita GDP has increased to more than US$ 2000 in 2014, up from approximately US$ 400 in the year 2000. And, remarkably, the Gini coefficient—a measure of wealth inequality—has remained consistently low, signaling that increased wealth is not being concentrated into fewer hands. Indeed, development in Vietnam appears to be positively benefitting stakeholders in both urban and rural areas.

One of the underlying factors of this initial success is that attention has been paid to rural communities during the reform process. For example, in the central province of Thừa Thiên-Huế, the introduction of commercial tree plantations, in particular *Acacia mangium*, *A. auriculiformis* and their hybrids, has provided opportunities for rural residents to enter into the economy by growing trees for the pulp and paper industry on land that has been allocated to them through the Forest and Land Allocation program. Incomes have increased, and one can see the evidence of this in the types of houses that have been built in recent years.

And yet, people in the rural districts of Thừa Thiên-Huế province still rely on natural forests and derive significant benefits from accessing the forest and harvesting products. For these people, it is critical to continue to access the forest in order to supplement incomes and diets with forest products. Communities still require rights over forests, and can contribute to their conservation and management. In fact, it is necessary to maintain, not only residents' rights over forest access, but also to retain their connection with the forest; a connection that can ultimately erode if, over generations, fewer and fewer people remain in contact with the natural environment.

Thus, on the one hand, economic progress is occurring rapidly and is benefitting the citizens of Vietnam. On the other hand, there is still a major sector of society that remains in close contact with forests, derives benefits from it, and should be integrated into long-term visions of conservation and management of resources. It was this complex framework that was the foundation for inquiry into multidisciplinary research focused on reconciling these somewhat conflicting goals.

The chapters in this third volume are the result of more than a decade of research and collaboration between the Asian Institute of Technology's Natural Resources Management field of study and the Hue University of Agriculture and Forestry (HUAF). Initial funding came in 2003 and 2006 from grants by the John D. and Catherine T. MacArthur Foundation's (USA) Conservation and Sustainable Development program. In 2009 the Ford Foundation's Hanoi office provided six years of funding to AIT to continue training and research collaboration between AIT and HUAF. Both foundations provided the necessary support for capacity building and conservation research, executed through postgraduate scholarships for key faculty members at HUAF and civil servants in Thừa Thiên-Huế province. The chapters herein reflect the individual efforts of the program collaborators as part of their MSc or PhD research.

The subject matter addressed in these chapters is quite varied, reflecting the complexity facing policymakers as they try to balance out the multiple intersecting agendas: conservation of biological diversity, community access to forests, gender-informed policy, conflict resolution, soil conservation, education, and species propagation. Although extensive, the research contained in this volume is not exhaustive. Much more effort is needed to continue to inform policymakers so that science-based planning, decision making, and policy can emerge that will promote continued economic growth, while safeguarding the natural heritage and ecosystem services provided by Vietnam's forests.

The theme that binds the research in this volume is that solutions to natural resources management dilemmas require governance, livelihoods, conservation, and management dimensions. In our research group, this theme has its intellectual roots in The Ostrom Workshop at Indiana University (previously called the Workshop in Political Theory and Policy Analysis). Under Professor Elinor Ostrom's leadership, interdisciplinary research tools to study communities and forests were developed by the International Forest Resources and Institutions (IFRI) research network. Professor Ganesh Shivakoti and I had the privilege of participating in IFRI trainings, which provided us with the theoretical and practical foundations for establishing an IFRI Collaborating Research Centre at the Asian Institute of Technology, Thailand.

This volume, therefore, represents not only the direct research outcomes of the authors, but also the intellectual heritage that can be traced through contributing faculty members at AIT to the interdisciplinary theoretical foundations pioneered by Professor Elinor Ostrom. It is hoped by all of us—authors and colleagues alike—that this volume will inform policy makers, practitioners, and researchers as they seek solutions to the complex problems associated with conservation and maintenance of forest resources, while at the same time continuing to improve livelihoods through economic opportunity. If it does, then it will be a successful bridging of academic research with practical implementation; a goal that has been a central motivation of Elinor Ostrom and those of us who have worked with her through the past several decades.

E.L. Webb
Department of Biological Sciences
National University of Singapore

Preface

During the last one and a half decades, we have learned numerous modules concerned with the understanding of major issues related to the management of natural resources. For example, speculative issues linked to the management of the environment and of natural resources are important for enhancing our understanding of the mechanisms of managing natural resource bases in the regions, as well as how different stakeholders interact with each other in managing natural resources. Similarly, we have been able to summarize the ways in which institutions function, and to adjust over time by observing these institutions, whereby the individual choices made render consequences which are due to these particular choices. This is one aspect of a "multi-level conceptual map" which offers a study of a specific hierarchical section of interactions made in a system.

Issues concerning natural resources are not static, neither are methodological advances in analyzing and studying dynamism. Central to the informed policy formation process in natural resource management, it has become possible to capture the possible status of resources, institutional arrangements, and structural complexities and their outcomes at diverse spatial and temporal scales. In recent years, significant advances in the scale of research inquiry have changed, and such methods have generated growth in environmental and livelihood outcomes of significance to key policy.

Similarly, studies put forward by the MacArthur foundations have embarked on long training programs to aid communities in managing the forest by exploring ways to protect the forest of central Vietnam, and in supporting forest dependent communities through long-term monitoring on biodiversity and domesticating valuable plant species. Continuing these efforts of the MacArthur Foundation, the Ford Foundation country office in Hanoi supplemented grants to strengthen knowledge at the graduate level on capacity building, and provided training workshops to provincial and local-level forest authorities and university instructors for further training on forestry authorities in Central Vietnam.

As a single blue-print solution is not enough to improve technological and managerial advances in resource management, research priorities, and methodologies have changing socio-political settings. These advances have helped policy makers pursue and prioritize development actions, which are based on unique case studies that have been reported in this volume for Central Vietnam mostly in regards to natural resource issues and effective management approaches. These include: (a) Land use changes and influences on forest conservations and livelihoods after implementation of the forest land allocation mechanisms. (b) Issues concerned with forest and biodiversity loss, and governance issues in Central Vietnam. (c) Climate change impacts among local communities, and ways of adapting to these changes. (d) Concerns about reconciling science with traditional practices in natural resource management. (e) Institutions, policy arenas, and decentralization of forest management. (f) Opportunities of payment for environmental services and pilot Reducing Emissions from Deforestation and Forest Degradation (REDD) + Measurement, Reporting and Verification (MRV) for future emission trading in retrospect. Though the case studies are from Central Vietnam, the issues and problems talked about, in general, replicate those present at a global scale, with a particular similarity to Africa and Latin America.

The chapters presented in this volume have received research support from the Ford Foundation country office in Hanoi, the Hue University of Agriculture and Forestry (HUAF) in its capacity building and concerted knowledge sharing pursuit in integrated natural resources management (INRM), and the Asian Institute of Technology (AIT). This particular collaboration between HUAF and AIT was aimed to make faculty members aware on theoretical understanding on policy and practical ongoing complex policy environments. Through this understanding, those common pool resources which have collective action dilemma would be easily understood and

easily managed. This was further supported by number of field research conducted in collaboration between AIT and HUAF. This notion was the basic foundation for undertaking joint graduate program in INRM. The process of exchanging knowledge was academically strengthened by collaborative training programs, such as by the International Forestry Resources and Institutions (IFRI) with the Ostrom Workshop in Political Theory and Policy Analysis at Indiana University. A network was also developed through participation in the International Association for the Study of Commons (IASC); the outcome of which was this volume.

Worsak Kanok-Nukulchai
President
Asian Institute of Technology
Bangkok, Thailand

Le Van An
Rector
Hue University of Agriculture
and Forestry
Hue City, Vietnam

INTRODUCTION

Natural Resources Dynamism and Management Concerns in Central Vietnam

G. Shivakoti*,†, S. Sharma‡, D. Hulse§, N.T. Dung¶, T.N. Thang¶

*The University of Tokyo, Tokyo, Japan †Asian Institute of Technology, Bangkok, Thailand ‡WWF Nepal, Kathmandu, Nepal §Ford Foundation, Jakarta, Indonesia ¶Hue University of Agriculture and Forestry, Hue, Vietnam

1.1 OVERVIEW

Central Vietnam is among one of the most dynamic regions of Southeast Asia. The fundamentals of the political and socioeconomic setting have been altered following the financial and economic turmoil in the region. Central Vietnam has observed 20 years of economic growth, and Vietnam is currently at "lower-middle-income country" status (World Bank, 2011). The economic growth, infrastructure development, and industrialization are swelling impacts on natural resources in the form of resources degradation and social turmoil at many stances (World Bank, 2011). The basic natural resources bases are decreasing at the cost to produce economic output. In a way, a part of these challenges have been offset by enhancing natural resource use efficiency and technology extension. However, the net end results are prominent in terms of increasing resources depletion and social unrest. Furthermore, climate change impacts have demanded further need for adaptation and mitigation measures to the consequences of erratic precipitation and temperature fluctuations, salt intrusions, and sea level increases that ultimately affect the livelihood of natural resource-dependent communities.

1.1.1 Socioeconomic Crises and Natural Resources Nexus in Central Vietnam

Particularly in Central Vietnam, ever since reunification took place during 1975, land was relocated to shifting cultivators to promote permanent settlements through transferring utilization rights to state forest enterprises. Natural resources management (NRM) is not just a technology fix and panacea but also helps in understanding the nuts and bolts of local institutions, their specific contexts, livelihood dependency, and opportunities of NRM. Local communities lost their traditional rights to the forest and were compelled to move away from swiddening to settle permanent agriculture, but they still practiced swiddening in a way that is not sustainable and further gave rise to environmental issues. The government of Vietnam had also developed a socioeconomic development plan (SEDP) that details the country's strategy en route to becoming a middle-income country on the foundation of a socialist market economy. This growth has been constructed on the foundation of a successful economic renovation mechanism (*doi moi*) that was initiated during the 1980s with poverty reduction as its primary emphasis. SEDP is focused on strengthening the business environment, gender and social inclusion, NRM, and environmental governance. By the end of the 20th century, the dependency on markets, building a multistakeholder economy, and *globalization* with regions and the world was rising through high dependency on a natural resource base. In Central Vietnam the population is increasing at an alarming rate followed by alternation of employment and settlements. Urbanization is slowly increasing with a rural-urban interface; although agriculture's contribution to the country's gross domestic product is reducing, the share from forest-based industries is increasing concomitantly as the number of standing trees is decreasing. A different and significant socioeconomic vitality that occurred in Vietnam is the decentralized decision-making process. There has been devolution of decision-making power to provincial and local governments for effective planning and service delivery. But studies show that counterpoint toward decentralization in Central Vietnam has occurred in terms of the inability of local government to extract benefits from larger resource while they may also promote an overblown setting up of a nationally important resource industry for their own benefits.

Protected areas in Central Vietnam were established during the 1960s through the 1980s for the protection of biodiversity. Nevertheless, deforestation and forest degradation and coastal mangrove depredation became rampant in these years both within and across protected areas and the entire landscape. This was enough for government to change their policy focus en route to conservation ethics and a broader emphasis on economic development. Buffer zones were then proposed as contrivances to defend protected areas from haphazard exploitation by local populations. But questions about the overall objectives of a buffer zone, authority for a management initiative, and ways to incorporate local participation to contribute to both protected ways and buffer zone management, all lacked clarity. Realizing the existence of all of these issues and prioritizing the importance of power devolution and local participation, a forestland allocation (FLA) policy was introduced that provided legal rights to people for managing the particular defined area for more than 50 years. But issues of cultivating acacia and rubber trees as mono-crop have been subject to discussions among environmentalists. Very little literature on livelihood and environment postimpact of FLA has been published to date, while learning from northern Vietnam has shown a negative evaluation.

Central Vietnam, following the expansion of the Ho Chi Minh trail to the national highway that connects Hanoi to Ho Chi Minh City, is at the center of discussions among conservation and environmental organizations as the route passes through biodiversity rich landscapes. The Central Highlands, home to some of the world's rarest species, has been bisected posing further threats to extinction. According to WWF (2007), highway construction has clearly resulted in soil erosion that has damaged surrounding forest area. Additionally, this has then impacted surrounding local rivers through clogged waterways and declining fish stocks. With the highway construction and hasty urbanization increasing, it has become easier for hunters to reach animals in remote areas. Animals often migrate for great distances to find mates and food, but the literature shows that infrastructure development has "lowered the space for free roaming and blocked wildlife corridors, fragmented habitats and pushed species towards extinction." Although the highway was intended to improve socioeconomic growth through better market access and establishing strong rural communication, its construction has speeded up cultural identity and livelihood loss of 37 ethnic minorities (WWF, 2007). Environment and social concerns were little considered during the design and planning of this construction project and this omission is responsible for major social, environmental, and political turmoil in Central Vietnam. In particular, shoreline areas have an intrusion of both freshwater and seawater from a distinctly brackish water ecosystem with high biodiversity, which forms an integral system of livelihood security for fishing communities. Ever since rapid sociodevelopment was undertaken, coastal urbanization has increased concomitantly affecting the overall ecosystem. There have been incidences of soil and land degradations, accelerated by climate change. As in many other developing countries, Vietnam needs a great amount of energy for its development and meeting targets conferred by its master plans. The demand for electricity has grown as other sources of energy are less developed. Especially, forests were destroyed during the hydropower boom from 2001 to 2010. According to informal statistics, there are more than 268 hydropower projects in Central Vietnam while another 205 are under construction; this means the deforestation scale is very high.

Together with infrastructure development, there are other socioeconomic development programs that equally influence the dynamics of natural resource use in the region. The national targeted program on New Rural Development (NRD) (2011–20) has forwarded 19 criteria for developing rural communes sustainably, which includes poverty alleviation, income generation, and socioeconomic enhancement. Along with this, the nexus between socioeconomic and environmental criteria is incorporated at local communes to enhance participation and foresee informed decision-making. In terms of forest conservation and local livelihood improvement, Decree 99/2010 has brought a new financing mechanism for forest conservation through the payments for forest environmental services (PFES) program. PFES in Vietnam has generated more than US$50 million annually. 85% of which has been paid to local communities to protect forest areas.

The Cancun Agreements ensures Vietnam will incorporate climate mitigation measures and development goals together with integrating the abovementioned issues through future Reducing Emissions from Deforestation and Forest Degradation (REDD+) programs. Earlier experience of successful implementation Payment for Environmental Services (PES) and *doi moi* could aid in assigning "property rights, transaction rules, conflict resolution mechanism, community benefit sharing schemes, transparent government and people's participation in decision making." REDD+ calls for ethnic minority participation in the policy and strategic plan formulation process though a transparent informed decision-making process.

1.2 DECENTRALIZATION AND POWER DEVOLUTION FOR NATURAL RESOURCES MANAGEMENT IN CENTRAL VIETNAM

Authenticating the vital role of the forest to the people living in the vicinity of a national park and protected areas, the Vietnamese government has allocated forest to individuals to improve the status of both the forest and individuals' livelihoods. This is often taken as a successful economic renovation policy as it has reduced poverty quite dramatically. As per Tuyen (2010), by the end of 2009 approximately 8.2 million ha of degraded forest area has been transferred among forest management councils/boards, state forest companies, communities, organizations, and individuals. Through the program, a paradigm shift from centrally managed forest to the state is observed that has encouraged forest protection, and the rationale for involving local people in NRM has been fulfilled through providing formal resource use rights. Community forestry is assumed to be legitimately defined among organizations and communities and is believed to support poverty reduction as it provides economic motivation for forest-dependent people to participate in forest management. Many authors believe that cost sharing among communities in resource management is likely to leverage poverty reduction. Plantation areas have increased while the forest management capability of local stakeholders increased accordingly. Research conducted by diverse research institutions and individuals confer FLA to have improved forest cover and quality with long-term community benefits. To the contrary, FLA has failed in the uplands of Central Vietnam, as observed by diverse authors in terms of the minor contribution to a household's income and reduced free access to the forest. Few literatures have mentioned FLA to have eroded the traditional resource use and management system among local forest beneficiaries. There is a lack of knowledge on the level of FLA implementation and the extent of its impact on the livelihood of resource-poor people. The limited literature shows potential risks that a household is likely to face through undergoing FLA processes.

1.3 FOREST GOVERNANCE AND IMPLICATIONS FOR FUTURE REDD+ PROGRAMS

REDD+ is likely coordinated and managed by national governments, promoted by public and private actors, and involves cooperation with government agencies or through a mutual combination of both (Corbers and Schroeder, 2011). Forest deforestation and degradation creates challenges to existing norms, rules, and a multitude of policies at different governance levels, and troubles economic development especially in developing countries (Humphreys, 2006). REDD+ is rooted in a superior governance framework as deforestation and degradation is closely related to global changes (Biermann et al., 2009). The institutions and organizations that managed the forest are ill-equipped to overcome the challenges of global transformations. A system is required that can transcend national boundaries, interconnect different governance levels, and allow both traditional and modern policy actors to cooperate. Such a system emphasizes integration of both formal and informal rule-making mechanisms and actor linkages in every governance stage that steer toward adapting and mitigating to local and global environmental change (Corbers and Schroeder, 2011).

1.4 SETTING UP AN INTELLECTUAL FORUM BY THE FORD FOUNDATION

The Ford Foundation has funded capacity building and a concerted knowledge-sharing mechanism in integrated natural resources management (INRM) at Vietnam's Hanoi University of Agriculture (HUA) and Hue University of Agriculture and Forestry (HUAF), and the Asian Institute of Technology (AIT) for collaboration intended to assist curriculum development and generate a body of knowledge for a mutual learning environment in the form of masters and PhD fellowships. This was basically intended to help faculty participate in understanding theories and diverse policy arenas for understanding and managing common pool resources (CPRs) that have a collective action dilemma through field research on a collaborative mode between AIT, HUA, and HUAF, which laid the foundation for a joint graduate program in INRM. During the period, AIT faculty participated in teaching at HUA until its number of faculty finished their higher studies in AIT and other institutions of higher learning in INRM. The process of exchanging knowledge was further strengthened academically by collaborative training programs such as International Forestry Resources and Institutions (IFRI) with the Workshop in Political Theory and Policy Analysis at Indiana University, but also developed a network through participating in the International Association for the Study of Commons (IASC), the outcome of which was this volume as an important textbook on INRM covering theory, its application, and related case studies. HUA and HUAF have a full-fledged graduate program running and the graduates from the program are already involved in influencing the policy and implementing effective local-level governance and management projects for INRM in Central Vietnam in particular and Vietnam in general. This book is one of the outcomes of the Ford Foundation's involvement in Central Vietnam.

1.5 LESSONS LEARNED FROM THE PROGRAM

The numerous lessons that have been learned during the collaboration in understanding the major issues related to MNR are discussed below.

1.5.1 Theoretical and Intellectual Contribution of Nobel Laureate Elinor Ostrom in Pursuing Analysis of Policies and Institutions

The speculative issues linked to the management of environment and natural resources are presented in this volume to bring about an understanding of the mechanisms in managing a natural resources base in the regions and how different stakeholders interact with each other in managing the natural resources.

Elinor Ostrom received a Nobel Prize for proving how commons is vital to the world based on the theory of "tragedy of the commons," which focused private property and centralization as a way of protecting finite resources from depletion. Counterbalancing the later part, Ostrom identified eight design principles on how CPRs could be governed sustainably and equitably in a community.

Similarly, the Institutional Analysis and Development (IAD) framework summarizes the ways institutions function and adjust over time. The framework observes that institutions are created by humans whereby individual choices made render consequences to particular choices made. This is part of a "multilevel conceptual map" that may offer to study a specific hierarchical section of interactions made in a system. The part of the framework includes action arena identification, formed through interactions between actors and actor situations.

1.5.2 Methodological Advancements and Diversification in Natural Resources Management

Issues in natural resources are not static, so there are methodological advancements in analyzing and studying the dynamism. An informed policy formation process in NRM is possible through the use of information that clearly captures the possible status of resources and the institutional arrangements, structural complexities, and outcomes at diverse spatial and temporal scales. In recent years, significant advances in the scale of research inquiry have changed and such methods have generated growth in environmental and livelihood outcomes of key policy significance.

The framework initiated by IFRI could be assumed to be a pioneer structure providing methodological guidance to a microlevel investigating relationship between forest users, local institutions, and forest conditions that were developed through case studies from different countries. This method facilitates collection of both ecological data and socioeconomic and institutional information among the forest-dependent communities and can aid in conducting baselines, analyzing over time the resource and governance changes, and most importantly sharing the information results with respective stakeholders and pertinent colleagues. Usually, problems in comparative research often arise due to diverse categories of data sets available and different methodologies applied. It is unique in a way that IFRI networks and research centers apply a common framework for research inquiry that is composed of variables from both science and society. IFRI instruments are formed on the basis of an IAD outline and theoretical and pragmatic studies conducted earlier.

Similarly, studies were put forward by the MacArthur Foundation for exploration of ways to protect the forests of Central Vietnam and to support forest-dependent communities through long-term monitoring on biodiversity, domesticating valuable plant species, and embarking on long training programs to aid communities in managing the forest. Like the Ford Foundation, the MacArthur Foundation provided a grant to strengthen knowledge among graduate-level capacity building and a training workshop for provincial and local-level forest authorities and university instructors for further training of forestry authorities in Central Vietnam.

And most recently, attempts to examine prototype design guidelines for multilevel forest governance across Asia for evaluating enforcement mechanisms have tested issues of cross-scale interactions (Makato and Shivakoti, 2015). A unique concern was how resource dependence and decision autonomy in respect to external environment and service providers could be related to forest management. In addition, distinctive ideologies on how effective forest governance could be intermediated through bridging multilevel outcomes were explored. Similarly, for natural resources that are scattered and spatially diverse, with different economic development contexts, spatial information advancement in terms of GIS-remote sensing for change detection and scenario analysis has recently been the foundation for efficient resource management.

A single blueprint solution is not enough to improve technological and managerial advances in resource management; research priorities and methodologies are dynamic with changing sociopolitical settings. These advances have helped policy makers pursue and prioritize development actions. In this context, four volumes (two each from Indonesia and Vietnam) are provided in this series to discuss the concerns and issues experienced recently.

1.5.3 Case Studies in Central Vietnam

Unique case studies are reported in this volume for Central Vietnam, mostly in regard to natural resources issues and effective management approaches. These include (a) land-use changes and influences on forest conservation and livelihoods as post-intervention impacts of FLA mechanisms; (b) issues concerning forest and biodiversity loss and governance in Central Vietnam; (c) climate change impacts and the existing body of knowledge among local communities and ways of adapting to these changes; (d) concerns of reconciling science with traditional practices in NRM; (e) institutions, policy arenas, and decentralization of forest management; and (f) opportunities of PES and pilot REDD+ for future emission trading in retrospect.

Although the case studies are from Central Vietnam, the issues and problems talked over in general replicate those existing at a global scale, with particular similarity to Africa and Latin America.

1.5.4 Link Between Northern and Central Vietnam

The Red River, the origination of Vietnamese civilization, initiates through northern Vietnam and much of the culture in Central Vietnam has been influenced by northern Vietnam. But, these two sections of Vietnam acted separately each with its own government. Though later reunited, cultural and linguistic differences delineate the regions distinctively and separate each territory. All the same, all of Vietnam rests in the tropics, and northern Vietnam has a humid subtropical climate while some belts of Central Vietnam have savannah grasslands. Based on the cultural and climatic differences, the natural resources and management interventions in these two parts of Vietnam tend to differ significantly. The issues, needs, and priorities tend to differ simultaneously. Volume II of this four-book series explicitly deals with resource and management dynamics concerns of northern Vietnam, while Volume III depicts the same in Central Vietnam. Mutual learning from these volumes may aid in applying concepts in their respective ways. These volumes may link these two regions to building problem-solving typologies to support, encourage, and coordinate one another to solve overall issues in NRM. This also encourages opportunities for learning that influence policy in both national and regional arenas.

1.6 RESOURCEWISE LOCAL MANAGEMENT PROBLEMS

1.6.1 Forests

Due to its varied climate, environments, and complex landscape, Vietnam has a variety of plants and soil types. During 1943 to 1990 a lot of Vietnamese forests were degraded; nevertheless, forest area increased from 2000 to 2005. Some scholars confirm Vietnam is going in the right direction, while others infer it as a result of plantation forests without due considerations to biodiversity. As such, traditional species that render a good market price are on the verse of extinction; monocropping through acacia and rubber plantations has severely aggravated the process. There have been incidences of deforestation for conversion of naturally existing forests to plantation crops. In Vietnam the forest belongs to the people, while the government is just a representative. But the representative government and the local community's mutually exclusive goals for preservation and use have led to serious competition and conflict over land allocation. The natural resources in the region are then overexploited and utilized in an unsustainable manner, which results in changed quantity, quality, and distribution of the natural capital.

1.6.2 Water Resources

The fishing communities and subsistence farmers are the most affected parties from climate-induced erratic rainfall and temperature phenomena arising as incidences of storms, hurricanes, and landslides. They have been experiencing silt depositions in the farmlands and fish mortality due to polluted water and siltation. Some fishing communities also believe these events, explicitly floods, bring in new fish species while others believe these events damage their fishing gear. Either way, natural hazards exert a strong influence on their livelihoods.

1.6.3 Land Resources

The Vietnamese people depend on forest and agriculture built upon land, and they understand the significance of land resources. Government attempts to resettle nomadic farming communities through FLA mechanisms have resulted in some negative consequences, as mentioned in the literature. FLA is believed to have created inequality among diverse forest dwellers, whereby resource-poor households tend to sell the available land with no capacity to invest in it. This book intends to explore the queries and impacts of FLA on people's livelihood.

1.7 BRIEF OUTLINE AND SUMMARY OF ISSUES ADDRESSED IN THE BOOK

Basically, there are 16 chapters in this volume, further divided into six sections depending on the similarity of issues being addressed: (a) Section I: Introductory; (b) Section II: Land-Use Systems and Livelihood Complexities; (c) Section III: Science and Natural Resources Management; (d) Section IV: Merging Science and Traditional Practices in Natural Resource

Management; (e) Section V: Institutions and Policy Dimensions; (f) Section VI: Concluding Section. A basic summary of a few of these sections is included below.

1.7.1 Section II: Land-use Systems and Livelihood Complexities

Though government has handed over the degraded land to the community through means of FLA schemes, scholars show that the tendency of converting degraded forest to tree plantation poses serious threats of biodiversity losses through monocropping. But contradictorily, 17 years after FLA implementation in Central Vietnam, apart from the fact that forestland had increased forest area and created awareness and supported livelihood of the resource-poor, questions on "which extent and level of FLA implementation have impacted on the livelihoods of the poor people, or what are the current and potential risks that the poor households have to face after FLA" not explored earlier is discussed in this section.

1.7.2 Section III: Science and Natural Resources Management/Section IV: Merging Science and Traditional Practices in Natural Resource Management

In the midst of forest being degraded at a faster rate, detailed scientific documentation of forest originality and existing vegetation is lacking. Explicitly, diverse vegetation occurring in the lowlands of Central Vietnam lack details on forest densities, floral richness, and abundances. Much of the targeted traditional species are on the verge of extinction due to knowledge lapses about its occurrence and cultivation practices. It is often observed that acacia and rubber are the primary species planted by the farmers, but lack of knowledge on plantation techniques and physiology of traditional species that are more valuable than these crops are least-favored by the locals. This situation of species extinction is aggravated by climate change. "Water" is one of the primary affected sectors of climate change. Past attempts are limited to understanding climate change causes and impacts in terms of precipitation fluctuation. But fishing communities, whose livelihood is in proximity with water sources, are under serious threat from changing climate. Often, studies conducted earlier focus on proposing solutions to help people adapt to climate change, while few of these "ways out" tend to work. A number of interventions must have been adapted by the communities themselves, and these could be documented and mainstreamed into the government's official plan. The issues do not rest here; the traditional way of considering/identifying forest conditions through the emergence of indicator species has been quite effective in resource management, primarily through successful resource management undertaken by traditional local forest beneficiaries. Ecologists and conservation practitioners have used this method of demarcating forest conditions through indicator species, particularly thorough an individual taxonomical group's emergence in a defined area. Researches have continued to improve the process, though this has been dynamic due to fluctuating climatic parameters. These issues are heavily addressed in this volume.

1.7.3 Section V: Institutions and Policy Dimensions

Prior to being reallocated as protected area, national parks in Central Vietnam were open access areas for locals to freely access forest products. A situation of conflict arose between the local forest-dependent communities and park officials on resource use and management. Having no other options, the locals started practicing intense shifting cultivation that was not scientific, undertook illegal poaching and tree felling, overexploited nontimber forest products, carried out incidences of forest fires, and made encroachments for agriculture expansions. All these activities had a negative impact on overall resource conditions. Along with Ho Chi Minh highway, Central Vietnam is home to a number of hydropower projects and infrastructure developments built with the objectives of energy production, agriculture drainage, and flood management. Prior to getting any construction underway, a large number of people are resettled and most of the time this may be difficult. In the context of hurdles between development, conservation, and livelihood nexus, these resettlement programs tend to be unsuccessful. With the view of fulfilling the basic life necessities, these individuals would be bound to exploit resources in a way that is not sustainable and pose the situation of future confrontation with the state. Vietnam, though chosen for the UN-REDD program, has developed strategies and plans to reduce drivers of deforestation but, mostly, these priorities often do not consider local needs and aspirations. This section has presented a brief overview of overall governance issues involved with REDD+ and learning from PES undertaken in terms of benefit sharing.

References

Biermann, F., Betsill, M., Gupta, J., Kanie, N., Lebel, L., Liverman, D., Schroeder, H., Siebenhuener, B., 2009. Earth System Governance: People, Places and the Planet. Science and Implementation: Plan of the Earth System Governance Project. Earth System Governance Report 1, IHDP Report 20. IHDP, Bonn.

Corbers, E., Schroeder, H., 2011. Governing and implementing REDD+. Environ. Sci. Policy 14, 89–99.

Humphreys, D., 2006. Deforestation and the Crisis of Global Governance. Earthscan, London.

Makato, I., Shivakoti, G. (Eds.), 2015. Multi-level Forest Governance in Asia: Concepts, Challenges and the Way Forward. Sage Publications, New Delhi/California/London/Singapore.

Tuyen, V.D., 2010. Real Situation and Solutions for Forestland and Forest Allocation in Vietnam. (Workshop Document). Ministry of Agricultural and Rural Development, Hanoi.

World Bank, 2011. Vietnam development report: natural resources management. Joint Development Partner Report to the Vietnam Consultative Group Meeting Hanoi, December 7–8, 2010.

WWF, 2007. Ho Chi Minh Highway Found to Have Mixed Impacts on Central Vietnam's Development. Retrieved from: http://wwf.panda.org/?114480/Ho-Chi-Minh-Highway-found-to-have-mixed-impacts-on-central-Vietnams-development.

SECTION II

LAND USE SYSTEM AND LIVELIHOOD COMPLEXITIES

Forest Conservation and Land-Use Change: A Case Study From a Remote Central Vietnamese District

Thiha

The Treedom Group, Bangkok, Thailand

2.1 INTRODUCTION

Vietnam's Central Truong Son Landscape (CTSL), one of the most important areas for biodiversity conservation, was highly forested throughout the 20th century. Over the last few decades, its forest cover has been considerably reduced and fragmented thanks to infrastructure development, logging, and clearance of forests for agriculture and overexploitation of nontimber forest products (NTFPs). This was the first national-scale forest loss that occurred during the past century, prompting the national government to implement several policies to conserve remaining pristine forests since the early 2000s. At the same time, improved infrastructural networks, agricultural technologies, and opening up local and regional markets resulting from major socioeconomic investments by the national government and international development agencies have motivated farmers to switch their land uses from subsistence farming to commercial agriculture.

This chapter addresses two related issues surrounding land-use change and forest conservation in Central Vietnam since the early 2000s: accelerated forest degradation events since forestland allocation (FLA) came into effect in Central Vietnam and rapid conversion of degraded forests and agricultural lands to tree plantations of rubber and acacia. Using a narrative approach, the chapter first analyzes the historical policy landmarks and their implications to the forest cover and land-use dynamics of Central Vietnam. The chapter then analyzes the factors influencing the change in farmers' land-use decisions toward tree plantations since the early 2000s using statistical modeling of household survey data in a remote central Vietnamese district. The chapter concludes with a forum of recommended actions that are aimed at stabilizing rapid forest degradation, easing agriculture pressure on forests, and sustainable management of forest resources.

2.1.1 Brief History of Forest Management in Vietnam

Centralization of forest resources management in Vietnam began in the late 1950s through the implementation of several national forest policies that mandated direct state involvement in the management, utilization, processing, and distribution of forest resources (Chung et al., 1998; Sikor and Apel, 1998). In the early 1960s, the national government transferred the management and use rights of forested land with a slope over 25 degrees to state forest enterprises (SFEs) and management boards (Minh and Warfvinge, 2002; Ba et al., 2003). In the late 1960s, the national government promulgated sedentarization policies to utilize land for food and industrial crops cultivation through village resettlements (Sikor and Apel, 1998; Müller and Zeller, 2002).

After national reunification in 1975, provincial governments in Central Vietnam began relocating swidden cultivators into permanent settlements located along forest edges at lower elevations, and encouraged them to adopt permanent use of land (ie, stable agriculture). Concurrently, provincial governments transferred the management and utilization rights of natural forests to SFEs and management boards, effectively terminating private ownership over forest resources (Tuan, 2005). As a result, local communities lost traditional rights over forests and were required to continue shifting away from swiddening toward permanent agriculture, although swiddening remained part of the agricultural system through the early 2000s (Tuan, 2005; Thiha et al., 2007).

In 1986, the national government launched *doi moi*, a policy of renovation, designed to implement a market economy guided by the national sector (Ari, 1998; Sikor and Apel, 1998; Müller and Zeller, 2002; Tuan, 2005). Trade liberalizations and integration into the international economy led to major changes in agricultural strategies (Litchfield et al., 2003). In Central Vietnam, the availability of new technologies (eg, high-yielding rice varieties, modernized processing techniques), credits/loans, and extension services from the Department of Agriculture and Rural Development (DARD) motivated farmers to focus on agricultural intensification (Müller and Zeller, 2002). In the early 1990s, the national government launched several forest decentralization policies that mandated the transfer of forest management authority to the provincial and district levels toward development of the forestry sector (Sikor and Apel, 1998; Sikor, 2001; Tuan, 2005). One of the cornerstones of decentralization policies was degraded forestland allocation (DFLA). Degraded forestland was either degraded forest or barren land earmarked for restoration of tree cover. Allocation refers to the process of providing legal, documented rights to specific households for protection over a defined area of land for 50 years for forestry purposes (Thiha et al., 2007). Along with the DFLA, there were major investments in plantation forestry by SFEs and local companies on allocated lands, so households could gain significant financial benefits from sales of trees to pulp mills (Thiha et al., 2007). Concurrently, the SFEs also signed short-term contracts with village households for natural forest protection, which included monitoring, patrolling, and forest fire protection, whereby participating households received nominal annual wages up to 50,000 Dong (~US$ 3) per hectare (Minh and Warfvinge, 2002; Tuan, 2005).

Since 2003, provincial governments (ie, Provincial People's Committees) in Central Vietnam began withdrawing some natural forests under the management of SFEs and allocating them to households and communities (Tuan, 2005). Under this natural forest allocation (NFA) policy, local governments became the implementing agencies of NFA and management.

2.2 METHODOLOGY

To understand the historical development of national forest policies and their linkage with forest-cover dynamics and change in farmers' land-use decisions toward tree plantations, a case study was conducted. This study used an integrated approach of data collection and analysis, which combined the qualitative assessment of historical policy landmarks during the period between 1975 and 2004 with a statistical modeling of household survey data.

In earlier 2004, Nam Dong District, located some 50 km south of Hue City in Thua Thien Hue Province, was selected (Fig. 2.1). Located at the latitude between 15°59.4′N and 16°14.4′N and the longitude between 107°30.6′E and 107°52.8′E, Nam Dong is one of the remote districts in Central Vietnam. It is a mountainous district with elevation varying between 40 and 1700 m above sea level. Covering an area of 650.5 km² across eleven communes (the smallest administrative unit in Vietnam), the landscape of the district is dissected by the tributaries of the Ta Trach River, forming many microcatchments (Thang, 2004).

2.2.1 Qualitative Analysis

The qualitative method combined a narrative description of historical policy landmarks with key informant interviews. Whereas narrative description synthesized the historical policy landmarks in three forest management periods (Table 2.1) by linking them with observed landscape outcomes qualitatively, key informant interviews focused the linkage between recent forest decentralization policy initiatives (ie, DFLA and NFA) and major socio-economic investments and farmers land-use responses.

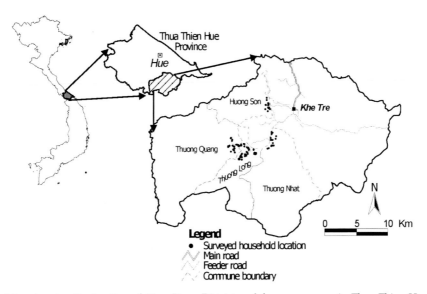

FIG. 2.1   Map showing the location of Nam Dong District and four communes in Thua Thien Hue Province, Central Vietnam, where a household survey was conducted (Thiha et al., 2007).

TABLE 2.1   Historical Policy Landmarks and Their Landscape Outcomes (1975–2004) in Nam Dong and Central Vietnam

Period	Dates, Policies, Decrees, and Documents	Description
1975–85 (State forest management)	1975, Sedentarization and fixed cultivation effective in Nam Dong, establishment of economic zones;	Relocation of swidden cultivators in permanent settlements located at forest edges at lower elevations and introduction of permanent use of land;
	1975, Transfer of forest rights to *Khe Tre* and Nam Dong SFEs, nationalization of forestlands, banning swidden cultivation, and termination of private ownership over forest resources.	Centralization of forest resources management, processing and utilization, demarcation of forestland with a slope over 25 degrees as the state property, increasing law enforcement on forest conservation.
1986–2000 (Market economy and devolution of forest resource management)	1986, *doi moi* renovation (in the economy) policy	Transition from a centrally planned economy controlled by state cooperatives to a market economy managed by individual households;
	1991, Law on Forest Protection and Development	Rules and regulations on forest management, protection, development, and utilization of forest resources;
	1993, The Land Law	Law on property rights of land, regulations on long-term user rights to individuals, households, groups of households, and organizations;
	1994, Decision No. 02/CP	Establishment of mechanisms and rules for allocating forest to organizations, households, and individuals for long-term (50 years) use for forestry purposes;
	1994, Decision No. 202/TTg	Contracting for forest protection and reforestation through participation of local people;
	1998, Decree No. 661/QD-TTg	National program of reforestation (Five million hectares reforestation project);
	1999, Decree No. 163/1999/CP	Implementation of DFLA: Mandate on forestland allocation and leases to organizations, households, and individuals for stable and long-term use for forestry purposes.
2001–4 (Socioeconomic development and NFA)	2001, Decision No. d178/2001/QD-TTg	Initiation of NFA: Stipulation on benefits and obligations of households and individuals allocated or leased or contracted forestland and forest;
	2003, NFA by Provincial People's Committee	Transfer of management rights over degraded natural forests to individuals, households, and communities for a 50-year period.

II. LAND USE SYSTEM AND LIVELIHOOD COMPLEXITIES

Key informant interviews were conducted with a total of 30 purposely selected respondents in Thuong Quang, Thuong Long, Huong Son, and Thuong Nhat communes of Nam Dong District. Key informants were comprised of 26 local residents including village headmen, heads of social associations, and older people with good knowledge of local history and four nonlocal residents including two staff each from the Nam Dong District Forest Protection Unit (DFPU) and the Nam Dong DARD office. Additionally, government statistics and relevant literature were reviewed to gain insights into farmers' land-use responses to the government's forest policies and local landscape outcomes.

2.2.2 Statistical Modeling of Household Land-Use Change

Factors affecting household land-use changes toward tree plantations since 2001 were investigated based on the analysis of the household survey data collected from 103 randomly selected households in four study communes in Nam Dong District (see Table 2.2). The household survey employed face-to-face interviews with respondents, most of them household heads, and was guided by a semistructured questionnaire, which was developed based on two reconnaissance surveys conducted earlier. The questionnaire was first pretested with 20 purposely selected respondents in four study communes and then translated into the Vietnamese language. Local field assistants used the translated questionnaire to guide their interviews with the respondents.

2.2.3 Method of Defining Household Land-Use Change

To understand the driving factors of major land-use changes between 2001 and 2004 (see Thiha et al., 2007), the dependent variable "conversion to tree plantations" was created. The variable represents the shift in the land-use choices of surveyed households, either from upland crops or natural forest, to tree plantations of rubber and acacia between 2001 and 2004. The variable was treated as a multinominal and its values were recorded for each survey household using the following rule: The variable had the value of "one" if the surveyed household did not convert its land that was registered as upland crops into tree plantations, "two" if the surveyed household converted any portion of its land that was registered as upland crops into tree plantations, and "three" if the surveyed household converted any size of

TABLE 2.2 Demographic Information of Four Study Communes and the Surveyed Households in Nam Dong District

Commune	No. of Villages	No. of Total Households (HHs)	No. of People	No. of Surveyed HHs			% of Total HHs
				Katu	*Kinh*	Total	
Huong Son	7	208	1251	16	0	16	7.7
Thuong Nhat	7	330	1733	17	1	18	5.5
Thuong Long	9	370	1232	39	4	43	11.6
Thuong Quang	7	283	1570	16	10	26	9.2
Total	*29*	*1191*	*6686*	*88*	*15*	*103*	*8.6*

natural forest into tree plantations since 2001. Because the variable was built on a temporal scale, it was necessary to assess the reliability of the information provided by the respondents using attribute data from field observations, such as age of the plantation, status of land parcels and surrounding land cover, and the choice of the surveyed households' land use prior to 2001.

Because the explanatory variables were generated from the data collected in 2004 through 2005, they were assumed to be invalid to explain household land-use changes toward tree plantations before 2001. Those cases where the surveyed households converted forest or land registered as upland crops into tree plantations of rubber and acacia before 2001 were therefore excluded from the statistical model.

2.2.4 Explanatory Variables of Household Land-Use Change

To understand the explanatory factors of household land-use change, a total of 14 explanatory variables representing household characteristics, socioeconomic status, physical accessibility, and policy were selected. Detailed descriptions of those variables are given in Table 2.3.

TABLE 2.3   Descriptive Statistics of the Explanatory Variables of Household Land-Use Change (2001–4)

Variable Name	Measurement Unit	Measurement Scale
Years since arrived (residential period)	Year	Continuous
Educational status of household head	1 = no formal, 2 = primary, 3 = secondary, 4 = higher than secondary	Ordinal
Household labor force	Number of people older than 14 years	Continuous
Total landholding	Square meter	Continuous
Landholding per capita	Square meter	Continuous
Annual per capita income	Million dong	Continuous
Membership in associations	1 = Nonmember, 2 = Member	Binary
Distance to paved road	Kilometer	Continuous
Time spent to reach paved road	Minute	Continuous
Distance to nearest local market	Kilometer	Continuous
Time spent to reach nearest local market	Minute	Continuous
Period of market used	Month	Continuous
Availability of extension service	1 = never, 2 = rarely, 3 = sometimes, 4 = always	Ordinal
Proportion of land owned under secure land tenure	1 = no tenure, 2 = less than a half, 3 = half, 4 = more than a half, 5 = completely secure	Ordinal

2.2.5 Variables Representing Household Characteristics

Five explanatory variables represented household characteristics of the surveyed households: ethnicity, years since arrival (residential period), household labor force, educational status of household head, and membership in associations (see Table 2.6). It was expected that households with a longer "residential period" would adopt new land uses earlier, as it would help them to accumulate experience on agricultural production and improve knowledge on capability of land and crop selection. Because they are at low risk of crops failure due to natural disasters (soil erosion, for example) and market price fluctuation in growing new crops, they tend to respond to the opening up of economic opportunities more quickly than newly arrived households. As a result, these households tend to change their land uses toward tree plantations more rapidly.

In Vietnam, persons older than 14 years are considered to be economically active, thus they are counted in the household labor force. "Household labor force" is an important input for agricultural production, which usually requires large labor supplies depending on the crop type and cropping intensity (Bao, 2005). On the one land, labor-rich households tend to embark on labor-demanding land uses (tree plantations, for example) to generate additional income. On the other hand, they might access forest more during the off-season than labor-poor households for the same purpose.

"Educational attainment" may help households consider environmental risks (soil erosion and crop damage, for example) in choosing particular land uses; thus, households with educated household heads likely choose permanent use of land (agroforestry-based, for example) that have minimal ecological impacts. Bao (2005) argued that educated households acquire better knowledge on the trade-off between current and future benefits/costs of a land use; thus, they prefer to adopt new land uses (tree plantations, for example) in pursuit of long-term economic goals. Also, educational attainment is a factor that discourages swidden cultivation practices and encourages permanent use of land (Müller, 2003).

Involvement in formal organizations and social associations is an important social attribute. At the household level, this can be measured by association membership of the household head. We expected that households with "membership in associations" (farmers' organizations, elders' associations, etc.) in place are aware of environmental risks associated with current and future land-use planning; thus, they prefer permanent use of land or land uses that have minimal ecological impacts.

2.2.6 Variables Representing Household Socioeconomic Status

Three explanatory variables represented the socioeconomic status of the surveyed households. "Income" is a direct, but important, indicator for measuring the affluence of a household. Households with a higher annual gross income should show how they depend less on forest than that with lower annual income by adopting particular land uses (tree plantations, for example).

"Size of landholding" is often a driving force of households to choose particular land-use strategies, because households with a larger landholding can afford more inputs and make improvements to increase its long-term productivity (Clay et al., 1994; Bao, 2005). They can also afford choosing particular land uses that usually require a larger area of land (rubber and acacia plantations, for example) to get higher economic returns. Because these households are comparatively wealthy, they can endure the short-term consequences from taking land out of production temporarily for particular uses; livestock or land restoration, for example.

"Period of market used" reflects how a farm household interacts with the local market in getting information about the market. It is expected that households that access the market much longer likely optimize their land uses earlier to maximize their productivity than those that access the market for a shorter time, as they are better informed about changes in market prices and demand for particular products available out of their land uses.

2.2.7 Physical Accessibility of Household

Four explanatory variables that measure the physical accessibility of surveyed households were selected: "distance to paved road," "time spent to reach paved road," "distance to nearest local market," and "time spent to reach nearest local market." It was expected that households with a better physical accessibility would choose more diversified land uses than that with poor accessibility, because diversified land uses may mean low financial risk resulting from changes in market demand and price of particular products available within their farmlands and the lower transportation cost may mean higher economic returns.

2.2.8 Policy Variables

A household's access to agriculture extension service is an important policy factor that influences household's land-use choice (Tuan, 2005; Bao, 2005). Since the early 2000s, farmers of Nam Dong were motivated to adopt new land uses, such as tree plantations, in pursuit of long-term economic benefits. In addition to the demonstrated profits from first rotation harvests of acacia plantations established in the mid-1990s in the district, availability of extension service from DARD provided farmers with sufficient incentives to convert their lands to tree plantations of acacia and rubber.

2.3 RESULTS

2.3.1 Landscape Dynamics of Nam Dong District (1975–85)

During this period of state forest management and village resettlement, three major landscape outcomes were recognized. First, establishment of economic zones in the wake of national reunification encouraged village households to embark on permanent use of land, resulting in a significant increase of irrigated agricultural lands in lower elevations. Second, nationalization of forestlands as part of the implementation of national forest protection policies prompted village households to abandon agricultural activities in natural forests, causing a significant decrease in shifting cultivation. Third, exclusion of local communities' participation in forest management and protection coupled with the transfer of forest rights to SFEs had effectively terminated communities' rights over use of forest resources and consequently created resource-use conflicts between local communities and SFEs. Because SFEs were required to fulfill the annual state timber demand, they began to intensify logging activities in the areas that were traditionally managed by local communities for subsistence livelihood. Loss of forest rights combined with emerging market opportunities and increasing population pressure prompted local people to access the forest under protection more rapidly, resulting in rapid degradation of dense forests.

2.3.2 Landscape Dynamics of Nam Dong After *Doi Moi* (1986–2000)

In the wake of promulgation of *doi moi* in 1986, the provincial government of Nam Dong began to relocate *Kinh* households, considered an ethnic majority, from coastal areas and the adjacent province of Quang Nam into the district's upland area. As of 2001, *Kinh* population in Nam Dong reached nearly 13,000 (Statistical Book of Nam Dong District, 2003), approximately a three-fold increase of its population size right after reunification. In the beginning of the relocation period, newly arrived *Kinh* households began to engage in large-scale land trading and land clearing as a means to solve the problems of land scarcity and consequent severe food insufficiency. At the same time, attractive revenue streams incurred from land sales and rentals motivated indigenous *Katu* households, considered an ethnic minority, who began to sell their lands and move further into the forest, where they cleared natural forests for subsistence agriculture.

In the early 1990s, the emergence of DFLA policies (Decision No. 02/CP in 1994, for example) coupled with a strong regional pulp and timber demand provided Nam Dong residents with the necessary incentives, prompting them to shift their land uses toward long-term cash crops (ie, rubber) and short-rotation tree plantations (ie, acacia and eucalyptus) more rapidly. Key informants reported that since the mid-1990s, farmers in Nam Dong and throughout Hue began to convert their lands earmarked for upland crops cultivation into tree plantations of rubber, acacia, and eucalyptus. In some cases where *Khe Tre* SFE signed plantation contracts with village households, farmers were allowed to cut degraded natural forests for the purpose of plantation establishment.

2.3.3 Landscape Dynamics of Nam Dong After the Year 2000

After 2001, forest degradation and land-use conversion toward rubber and short-rotation tree plantations began to accelerate. Those dynamic changes within forest and outside forest boundaries coincided with the implementation of NFA policies (Decision 178 in 2001, for example; Table 2.1) and major socioeconomic investments in the district. During this period, two interconnected proximate causes were found responsible for degradation events in natural forests. First, since the early 2000s, *Khe Tre* SFE began to intensify selective logging activities in natural forests that were earmarked for allocation to individual households or groups of households under NFA policies. Subsequently, local people gained increased access to the forest on the logging trails that were newly established or improved by the SFEs. Second, there were substantial delays in the NFA process mainly due to excessive bureaucracy and poor coordination among implementing agencies of NFA (Thiha et al., 2007).

Outside the forest boundaries, farmers of Nam Dong continued to expand the areas of rubber and acacia plantations due to the strong regional market demand for raw rubber and pulpwood, potentially high economic returns, and availability of attractive loans and subsidies from the SFEs (Thiha et al., 2007). Conversions of upland crop areas into tree plantations continued and substantially increased in the case of acacia plantations, as first rotation harvests occurred in the province and it became clear that profits from short-rotation tree plantations were exceptional.

At the same time, farmers in Nam Dong began to convert degraded forests into acacia and rubber plantations. The majority of conversions allegedly occurred in forests that were in the process of being allocated. Having been supported by the DFLA policy, farmers were allowed to cut up to 2 ha of degraded forests for the purpose of plantation establishment (Tuan, personal communication). Besides the availability of credits and loans from DARD, farmers' increased motivation toward conversion of natural forests into plantations was largely influenced by anticipated financial benefits from sales of timber that was cut (Thiha et al., 2007).

The NFA policy in 2001 presented an unintended incentive for residents of Nam Dong to convert degraded forest into plantations. Very strong market demand for forest products, combined with a slow rate of allocation and poor enforcement of forest conservation by government agencies resulted in a rapidly emerging trend of land speculation taking place in Nam Dong. Key informants reported that since the early 2000s, the SFE's logging activities in the forests earmarked for allocation sharply increased. At the same time, local people accessed forests more through logging trails. In some cases, farm households resumed shifting cultivation within natural forests that were in the process of allocation.

2.3.4 Analyzing the Factors Affecting Household Land-Use Change Toward Tree Plantations Since 2001

A correlation analysis with the explanatory variables revealed that there were high correlations among several variables. The number of explanatory variables was thus reduced using principle component analysis (PCA) to fulfill the criterion of constructing a statistical model with a full set of uncorrelated independent components or factors (Li and Yeh, 1998; Liu and Lathrop, 2002).

PCA returned six principle components, which explained 75.9% of the total variance observed in the original variables (Table 2.4). The extracted principle components were named after thorough interpretation of the loadings of the original variables corresponding to the principle components in the rotated component matrix (Table 2.5). Finally, component loadings (standardized) were saved as variables.

2.3.5 Factors Influencing Household Land-Use Change Toward Tree Plantations (2001–4)

The M-logit explaining the influence of the explanatory variables over household land-use change consisted of two equations, each calculating the logit for a land-use change type with reference to no change. The explanatory powers of the variables in each of the logits are given in Table 2.6. Of the total of six explanatory variables entered in the model, four variables did not contribute significantly to the model: physical accessibility, education membership in associations, land tenure, and market and labor. In each of those variables, the difference in -2log-likelihood with and without variable was not significant at 0.05 level (see Table 2.6).

TABLE 2.4 Extraction of the Principle Components From the Original Variables Using PCA. The value in bold phase represents the cumulative percentage of the loading scores of the first six components after rotation.

Component	Initial Eigenvalue			Extraction Sums of Square Loadings			Rotation Sums of Squared Loadings		
	Total	% of Variance	Cumulative %	Total	% of Variance	Cumulative %	Total	% of Variance	Cumulative %
1	3.37	24.08	24.08	3.37	24.08	24.08	2.98	21.27	21.27
2	2.57	18.33	42.41	2.57	18.33	42.41	2.31	16.52	37.79
3	1.39	9.91	52.32	1.39	9.91	52.32	1.62	11.54	49.33
4	1.16	8.27	60.59	1.16	8.27	60.59	1.28	9.11	58.44
5	1.11	7.93	68.52	1.11	7.93	68.52	1.23	8.79	67.23
6	1.04	7.42	75.95	1.04	7.42	75.95	1.22	8.71	**75.95**
7	0.95	6.79	82.74						
8	0.72	5.14	87.88						
9	0.58	4.18	92.05						
10	0.43	3.08	95.13						
11	0.38	2.75	97.88						
12	0.15	1.07	98.96						
13	0.09	0.61	99.57						
14	0.06	0.43	100						

II. LAND USE SYSTEM AND LIVELIHOOD COMPLEXITIES

TABLE 2.5   Rotated Component Matrix Using *Varimax* and *Kaiser* Normalization Rotation Method

Variable	Physical Accessibility	Landholding	Education and Membership	Residency Period	Tenure Security	Market and Labor
	C1	C2	C3	C4	C5	C6
Distance to paved road	**0.87**	−0.15	−0.09	0.18	−0.03	0.09
Time spent to reach paved road	**0.87**	−0.06	−0.13	0.03	−0.03	0.05
Distance to local market	**0.74**	0.42	−0.05	−0.13	0.30	−0.07
Time spent to nearest local market	**0.78**	0.34	−0.12	−0.15	0.26	−0.18
Total landholding	0.11	**0.92**	0.12	0.10	−0.03	0.10
Landholding per capita	0.03	**0.93**	0.02	−0.01	0.03	−0.15
Educational status of household head	−0.23	0.10	**0.69**	−0.09	0.31	0.16
Membership in association	−0.03	0.12	**0.70**	0.33	−0.05	−0.04
Availability of extension service	−0.15	−0.04	**0.67**	−0.36	−0.24	−0.16
Years since arrived	−0.01	0.03	−0.02	**0.86**	0.02	−0.06
Proportion of land under secure tenure	0.16	0.01	0.00	0.02	**0.88**	−0.10
Period of market used	−0.06	−0.13	−0.16	−0.24	0.01	**0.79**
Household labor force	0.09	0.16	0.21	0.37	−0.19	**0.63**
Per capita income	−0.42	0.47	0.26	0.07	0.32	0.25

Rotation converged in eight iterations. Only the loading scores in bold phase from the original variables were interpreted. C1 to C6 represent principle components.

TABLE 2.6   M-Logit Model for Household Land-Use Conversion Toward Tree Plantations (2001–4) With Reference to "No Change" ($n = 100$). The values in bold phase are considered statistically significant.

Variables (PC Scores)	Upland Crops to Plantations				Natural forest to plantations			
	B	Std. Error	Sig.	Exp (B)	B	Std. Error	Sig.	Exp (B)
Intercept	4.00	1.02	0.000		1.66	1.09	0.127	
Physical accessibility	0.45	0.48	0.351	1.56	0.85	0.61	0.162	2.34
Landholding	2.60	0.94	**0.006**	**13.47**	3.10	0.99	**0.002**	**22.18**
Education and association membership	−0.50	0.62	0.424	0.61	−0.66	0.69	0.336	0.51
Residency period	1.13	0.54	**0.035**	**3.10**	0.80	0.64	0.212	2.22
Land tenure	−0.78	0.56	0.165	0.46	−0.73	0.69	0.288	0.48
Market and labor	0.67	0.49	0.175	1.95	0.42	0.57	0.457	1.53
Model summary statistics								
Chi-square statistics (Likelihood ratio test; $df = 12$, $p = 0.00$)	40.93							
Cox and Snell R-square	0.34							
Nagelkerke R-square	0.45							
(McFadden)	0.3							
Percentage correctly predicted: No change	58.3%							
Upland crops to plantations:	94.9%							
Natural forest to plantations:	0%							
Overall	81%							

The significance test of the regression coefficients showed that in the first logit (ie, upland crops to plantations vs. no change), landholding size and residency period were significant at 0.01 and 0.05 levels, respectively. Landholding size had the highest explanatory power in the model that one unit (principal component score) increase in landholding size would increase the odds to convert upland crops to tree plantations by a factor of 13.5 and with reference to households that did not change their land uses during the 2001–4 period. Residential period was the second important variable associated with the conversion of upland crops to tree plantations since 2001 (see Table 2.6).

In the second logit (ie, natural forest to tree plantations vs. no change), landholding size was the only important predictor for the conversion of natural forest to tree plantations that one unit (principle component score) change in landholding size would increase the odds to convert natural forest into tree plantations by a factor of 22.2 compared with the reference category (see Table 2.6).

The likelihood ratio test showed that the model was significant at 0.05 level, rejecting the null model (ie, model without the explanatory variables). The test for goodness-of-fit showed that the model had a good fit to the empirical data set, as indicated by large R-squares values. Large R-square values also indicate that the strength of relationship between the explanatory variables and the odds for households to change their land uses was strong. The model had a good prediction power as well, as the choice of the household to convert upland crops to tree plantations was correctly predicted for 94.9% and overall prediction success of the model was as high as 81%. However, the model failed to predict for the choice to convert natural forest to tree plantations (see Table 2.6).

2.4 DISCUSSION

Narrative findings from synthesis of historical policy landmarks were consistent with that of an earlier quantitative land-cover change study in Nam Dong (see Thiha et al., 2007) and revealed that the majority of tree plantations arising between 2001 and 2004 were upland crops before 2001. This observed trend of rapid decline of upland crops area and accelerated expansion of tree plantations were also commonplace in northern Vietnam (see Sikor, 2001). Müller and Zeller (2002) argued that this rapid shift in farmers' land-use decisions toward commercial agriculture was largely supported by agriculture development policies in the wake of the *doi moi* renovation policy. In Nam Dong, a number of socioeconomic, policy, and market factors were found responsible for the observed trend of household land-use change toward tree plantations since the early 2000s.

Qualitative analysis described the link between the observed landscape outcomes and policy and socioeconomic driving forces in each study period. For example, key informants reported that state-sponsored logging activities and consequent small-scale timber cutting in the forests that were in the process of being allocated were directly responsible for the observed rapid forest degradation since 2001. Household land-use changes toward tree plantations since the early 2000s were linked with the availability of agricultural loans and extension services from state agencies and demonstrated profits from first rotation tree plantations in the province. Narratives also identified the underlying forces of forest degradation events, such as increased local and regional market demands for forest products, slowly implemented forest allocation (ie, NFA), endowment of forest rights under NFA (ie, Red Book) and weak law enforcement on forest protection. Collectively, all of them encouraged local households to access forests more and resume shifting cultivation in natural forests, leading to a highly dynamic local landscape in Nam Dong and Central Vietnam.

Rapid land-use changes outside forest boundaries, especially conversion of upland crops to tree plantations, were mainly due to demonstrated high economic returns from harvests of first rotation acacia plantations, availability of loans and extension services, and increased local and regional market demand for pulpwood and raw rubber. Conversion of degraded forests into tree plantations was linked with the availability of forest rights under plantation contracts with SFEs and land speculation by local households. In the long run, those driving forces may impose important threats to forest and biodiversity conservation in Central Vietnam.

Qualitative analysis also identified the underlying forces of forest degradation events since national reunification. Key informants described that intensified logging activities by the SFEs and reclaiming forests for agriculture by local households were the major driving forces behind a highly dynamic landscape of Nam Dong and Central Vietnam. Since the early 2000s, those human activities within forest boundaries began to intensify and coincided with the implementation of recent forest decentralization policies (ie, NFA) and major socioeconomic investments by provincial governments and international development agencies.

These narrative findings were consistent with a prior study that recent NFA polices induced rapid forest degradation in Nam Dong (see Tuan 2005). Since NFA came into effect in Nam Dong, *Khe Tre* SFE began to intensify logging activities in forests available for allocation. At the same time, local households began to access forests more to cut trees to generate extra income with the speculation that those forests could be allocated to them at a later date. In some cases, farmers reclaimed forests for upland crops cultivation.

Statistical modeling of household land-use change since 2001 revealed that landholding size was an important factor associated with farmers' decisions to convert upland crops to tree plantations. This was likely because larger landholding may mean higher potential for commercial agriculture with the adoption of new land uses. Households with larger landholding sizes are at low risk of crop and market failures in adopting new land uses (tree plantations for example), possibly encouraging them to diversify their land uses to generate extra revenue streams. Conversely, small-landholding households cannot afford diversifying their land uses for the same purpose, as they are at high risk of crop and market failures, possibly discouraging them to adopt new land uses.

Analysis of household survey data coupled with field observation suggest that the majority of small-landholding households were inclined to engage in subsistence agriculture and produce crops that allow them to purchase rice and staples. In contrast, households with a larger landholding tend to set certain portions of their lands aside for soil conservation or fertility restoration purposes. In the event of opening up local and regional markets for short-rotation crops, the availability of those reserved lands would provide them with unintended incentives to embark on new land uses.

Landholding size was also important for conversion of natural forest into tree plantations since 2001. Naturally, a plantation of larger size will provide a farm household with higher economic benefits than that of a smaller size will, controlling for the other factors, such as soil fertility, slope gradient, and inputs. Larger landholding will also allow farmers to perform major tending operations more effectively at minimal costs throughout the rotation to increase the volume and value of planted trees with the purpose of timber and pulpwood production. Aside from landholding size, residential period was another important factor for conversion of upland crops into tree plantations. Households with a longer residential period are endowed with an accumulated experience on agricultural production. This combined with better knowledge on the inherent capability of land, crops choice, and market demand would encourage them to adopt new land uses more rapidly.

Qualitative analysis described the link between those observed landscape outcomes but major policy landmarks also identified the underlying factors of those dynamic changes. Key informant interviews revealed that forest degradation events and rapid land-use transition toward tree plantations were linked with recent policy initiatives, especially NFA, that mandated the allocation of natural forest to individual households and groups of households. Internal forest transitions accelerated since 2001, coinciding with implementation of forest decentralization policies (ie, NFA) in Nam Dong. Degradation events observed in closed-canopy forest were due to intensified logging activities by SFEs, and consequently small-scale logging by individuals, who had improved forest access on logging trails. Illegal timber cutting and, in some cases, land reclaims for swidden cultivations in degraded forests were arguably landscape outcomes from NFA policies.

Degradation events in degraded forests that could be allocated at a later date accelerated after 2001. Narratives described that those degradation events were the result of signiticant levels of land speculation by, in most cases, indigenous *Katu* households and, in some cases, newly arrived *Kinh* households. At the underlying level, those increased human activities within degraded forests were allegedly supported by weak forest conservation policies, slowly implemented forest allocation, and growing market demand for selected forest products.

During the same period, major land-cover changes outside forests were due mainly to farmers' land-use responses mediated by increased awareness of economic potentials of long-term cash crops and tree plantations, and availability of loans and extension services from the state agencies. At the underlying level, those may have been underpinned by a complex set of technological, demographic, and policy factors. A few examples that are also commonplace in northern Vietnam include introduction of new technologies, change in market price and taxation, infrastructural development (road construction, for example), increased population pressure, and change in demographic composition.

2.5 RECOMMENDED ACTION POINTS

Findings of this case study suggest that SFEs should decrease state-sponsored logging activities to stabilize rapid forest degradation events, which may become an important threat to forest and biodiversity conservation in Central Vietnam in the long run. The local government (ie, Provincial People's Committee) should proceed with forest allocation more rapidly to provide local households with long-term forest rights, which will consequently serve as sufficient incentives that will encourage them to improve and protect the allocated forests under their management. Small-scale industries should be promoted to provide farmers with off-farm job opportunities, which will eventually reduce agriculture pressure on forests during the off-season. Production of NTFPs such as rattan and Malva nut tree (*Scaphium lychnophorum*) should be integrated into existing agroforestry systems located within allocated forests to increase the forest diversity and to generate long-term stable revenue streams. Besides those production-based alternatives, the enforcement of forest laws should be enhanced to protect degraded forests from further depletion. Current forest conservation policies should be amended to allow local people to play a more important role in decision making. Equally important is building capacities of the staff of provincial governments and organizations to strengthen coordination among them, thereby to assist in forest allocation process and forest conservation projects.

References

Anon, 2003. Statistical Book of Nam Dong District for the year 2003. In Vietnamese.

Ari, N., 1998. Vietnam's *Doi Moi* policy and forest protection: the possibility of people's participation. Unpublished document.

Ba, H.T., Uan, L.C., Quang, V.D., Mau, P.N., Lung, N.N., Dung, N.Q., 2003. People, land and resources in the Central Truong Son Landscape. Central Truong Son Initiative Report No. 5. WWF Indochina, Hanoi, Vietnam.

Bao, L.Q., 2005. Multi-agent systems for simulation of land-use and land cover change: a theoretical framework and its first implementation for an upland watershed in the Central Coast of Vietnam. Ecology and Development Series No. 29, Cuvillier Verlag, Göttingen, Germany.

Chung, V.T., Crystal, E., Dzung, N.H., Dzung V.V., Phon, N.H., Poffenberger, M., Sikor, T., Sowerwine, J., Walpole, P., 1998. Stewards of Vietnam's upland forests. Research Network Report, Vietnam.

Clay, D.C., Guizlo, M., Wallace, S., 1994. Population and land degradation. Working Paper No. 14, USAID-Funded Global Program, the Environmental and Natural Resources Policy and Training Project (EPAT).

Li, X., Yeh, A.G.O., 1998. Principle component analysis of stacked multi-temporal images for the monitoring of rapid urban expansion in the Pearl River Delta. Int. J. Remote Sens. 19 (8), 1501–1518.

Litchfield, J., McCulloch, N., Winters, L.A., 2003. Agricultural trade liberalization and poverty dynamics in three developing countries. Am. J. Agric. Econ. 85 (5), 1285–1291.

Liu, X., Lathrop, R.G., 2002. Urban change detection based on an artificial neural network. Int. J. Remote Sens. 23 (12), 2513–2518.

Minh, V.H., & Warfvinge, H., 2002. Issues in management of natural forests by households and local communities of three provinces in Vietnam: Hoa Binh, Nghe An, and Thua Thien-Hue. Asia Forest Network Working Paper Series 5.

Müller, D., 2003. Land-use Change in the Central Highlands of Vietnam: A spatial econometric model combining satellite imagery and village survey data. PhD Dissertation, Institute of Rural Development, Georg-August University of Göttingen.

Müller, D., Zeller, M., 2002. Land use dynamics in the central highlands of Vietnam: a spatial model combining village survey data with satellite imagery interpretation. Agric. Econ. 27, 333–354.

Sikor, T., 2001. The allocation of forestry land in Vietnam: did it cause the expansion of forests in the northwest? Forest Policy Econ. 2, 1–11.

Sikor, T., Apel, U., 1998. The possibilities for community forestry in Vietnam. Asia Forest Network Working Paper Series 1, 27 pp.

Thang, T.N., 2004. Forest use pattern and forest dependency in Nam Dong district, Thua Thien Hue province, Vietnam. (Master's Thesis). Asian Institute of Technology, Bangkok, Thailand.

Thiha, Webb, E.L., Honda, K., 2007. Biophysical and policy drivers of landscape change in a central Vietnamese district. Environ. Conserv. 34 (2), 164–172.

Tuan, H.H., 2005. Decentralization and local politics of forest management in Vietnam: a case study of Co Tu ethnic community. (Master's Thesis). Graduate School, Chiang Mai University, Thailand.

Impacts of Forestland Allocation on Livelihood Activities and Income of the Poor in the Upland Area of Vietnam

L. Van Lan*, G. Shivakoti[†,§], N.T.T. Lien[‡]

*Hue University of Agriculture and Forestry, Hue, Vietnam [†]The University of Tokyo, Tokyo, Japan [‡]Hue University of Sciences, Hue, Vietnam [§]Asian Institute of Technology, Bangkok, Thailand

3.1 INTRODUCTION

3.1.1 Background

Vietnam has a total of about 12.90 million hectares of forest and forestland with forest cover of 38.3%. The forest and forestland are categorized into three kinds depending on their functions: special-use forest and forestland is around 15.6%, protection forest and forestland is approximately 48.1%, and production forest and forestland is about 36.3%. Vietnamese forest and forestland concentrate mainly in upland areas which account for around 57% of the total natural land of the country (MARD, 2008). Upland areas also are the home of nearly 22% of the national population with different groups of ethnic minorities that have low education, an underdeveloped economy, and a difficult life (Cuc, 2002; Thuan, 2005). Therefore, it can be said that the forest and forestland play an important role in poverty reduction and sustainable development of these areas.

Recognizing the integral role of forest and forestland to upland people, since the year 1990, the Vietnamese state has allocated its forest and forestland for the organizations, individuals, households, and communities to use stably to improve people's livelihoods as well as to contribute to forest protection and development. Up to 2009, around 8.422 million hectares of forest and forestland have been allocated in the whole country, including (1) for forest management boards: 4,318,492 ha; (2) for the state forest companies: 2,044,252 ha;

http://dx.doi.org/10.1016/B978-0-12-805452-9.00003-5

(3) for units of the army: 243,689 ha; (4) for communities: 191,361 ha; (5) for households and individuals: 872,734 ha; and (6) for other units of the economy: 751,472 ha. At present, Commune People's Committees (CPCs) in the whole country have been managing over 2,442,485 ha of forest and forestland, which will be allocated for households and communities in the near future (Tuyen, 2010). Almost all of the forest and forestland allocated for individuals, households, and communities are protection and production types.

3.1.1.1 Rationale

After 17 years of the implementation of forestland allocation (FLA) in Vietnam since 1991, aside from the successes of the program such as the forestland has had owners, lots of local people have actively invested to intensify their allocated forestland yields, people's awareness in forest management and protection has been enhanced, and the level of forest cover has been increased quickly (Hoang and Son, 2008), many questions have been raised as to what extent and level FLA implementation has impacted the livelihoods of poor people, or what are the current and potential risks that poor households have to face after FLA. These issues will be discussed and analyzed in this chapter.

3.1.2 Research Objective and Questions

The objective of this research is to explore the impacts of the implementation of FLA to livelihood activities and income of the poor households in the upland area of Central Vietnam. From that, the research will give recommendations for better-allocated forestland based on the poor households' livelihoods.

To achieve the above objective, two questions must be answered, as below:

1. Have livelihood activities and income of the poor households changed compared to the time before FLA implementation? If yes, what are the main causes for these changes?
2. What are the current and potential risks of having any of the transforming of livelihood activities and income of the poor households?

3.1.3 The Literature

In Vietnam, allocation of land to households and individuals is one of the successes of the *doi moi* (renovation) policies. This has created great economic growth as well as dramatic poverty reduction since the middle of the 1980s. The Land Law of 1987 revised in 1993 and 2003 gave people the right to exchange, inherit, mortgage, lease, and transfer their allocated land (Sunderlin et al., 2005). Until 1999, there were about 10 million households and individuals who received Red Books, mainly in the lowlands (Ba, 2002). In this program, FLA contributed an important part because it is a radical policy shift to the devolution of forestland management from the state to lower levels. The objective of FLA is to encourage forest regeneration and protection in the uplands, and the rationale of decentralization is that local people would be more interested in natural resources management (NRM) if they had formal rights of natural resources use (Sikor, 2000).

Community forestry has legally been accepted in the Law of Forest Protection and Development revised in 2005. The law defines the terms and conditions for FLA not only for organizations, households, and individuals but also for communities. Community forestry could offer more support for poverty reduction because it provides a significant economic motivation for people to participate in forest protection and management. The transfer of decision-making powers from the state to communities could also be a basic foundation for the improvement of living standards (Sunderlin et al., 2005). Blockhaus and Vu Dung also think that poverty reduction could be leveraged due to sharing the costs and benefits among members of the communities (Thuan, 2005; Dung, 2006).

As a result of the implementation of the Forest Protection and Development Law, the forest and forestland that was allocated to communities in Vietnam has increased quickly. In 2004, local communities in the whole country were allocated more than 2 million hectares of forest and forestland for use and management. The areas of plantation forest and the rate of forest cover have increased quickly. The capacity of forest protection and management of local communities has significantly improved. Their living standards are improved also due to taking part in the program of development of home gardens and forest plantations (Vien et al., 2007).

A literature review on FLA and community forestry in Vietnam showed that they made both positive and negative impacts on the livelihoods of local people.

3.1.3.1 Positive Evaluation

Some researchers agreed that after FLA, the control over land and land use of local people increased in some areas, resulting in reforestation and greater benefits for them (Nhan, 1998; Sunderlin et al., 2005). Research conducted by the International Development Research Center (IDRC) showed that forest cover and its quality improved dramatically after implementing FLA to farmers for long-term management. Benefits from the management of forest include the security of water sources for agricultural production, intercropping products, and nontimber forest products (NTFPs) (Bellamy, 2000). In Bac Giang Province, after receiving the allocated forestland, local farmers started to establish forest plantations to better exploit the barren lands and barren hills. In this case, FLA not only helps local people to improve their income, but also supports them to overcome landlessness and poverty. After possessing the forestland, local people had a big sense of ownership. They channeled their land resource actively into renovating production methods, investing new technologies with appropriate tree species, and enhancing productivity as well as quality of agroforest products (Thanh, 2000). In some northern areas, many kinds of fruit trees have been planted to replace cassava on allocated forestland. This has helped local people to escape at least partially out of a poverty condition (Thu, 1999; Ngan and Tho, 2000). FLA in Daklak Province demonstrated one of the most progressive efforts in Vietnam in moving toward community forestry. Local people are allowed to harvest NTFPs and are entitled to 6% of the total value of timber for every year since they began to manage the forest (Thanh, 2001). In the buffer zone of Bach Ma National Park (BMNP), local farmers plant perennial trees to diversify their income sources as well as to enrich the forest with both acacia and indigenous species, and bamboo for shoots. They also establish small-scale nurseries for seedlings to grow on their allocated forestland. As a consequence, not only the livelihood of these cultivators is improved but also the illegal forest exploited pressure from buffer zone to the national park is reduced (Lan et al., 2002).

3.1.3.2 Negative Evaluation

Contrary to the abovementioned authors, a negative view of FLA emerges from other sources. The Social Forestry Development Program (SFDP) showed that both FLA and incentives for planting trees failed in many upland areas in Vietnam. Forestry production accounted for just a small part of cultivation systems and contributed a minor percentage of household income (Due, 1999;

ADB, 2000). Research in Bac Can Province revealed that allocation of forestland to individuals and households reduced (even put an end to) the free access to the forest; therefore, it created many difficulties for people, especially poor cultivators who traditionally migrated and relied on the slopping cultivation (Castella et al., 2004). Besides, Apel (2000) stated that FLA has brought disadvantages to the development of forest in the upland regions because it eroded the traditional NRM system of local inhabitants. Relating to the equity, FLA is often inequitable with forestland overallocation for mass organizations, the employees of state forest enterprises (SFEs), the rich dwellers, and the powerful people. Hence, forestland is becoming a capital accumulation for individuals and households who have close access to political power and social networks (ADB, 2001). Phuong (2004) and Sunderlin (2005) also stressed in their research that only barren lands and barren hills are allocated to poor households, while the woody lands are allocated to SFEs. According to Tuan (1999), Kinh inhabitants (majority people in Vietnam) and other influential powerful ethnic groups have had tended to get more benefits from FLA in contrast with the poor and ethnic minorities living in remote areas. As the poor and ethnic minority households do not have enough financial resources to invest for afforestation, they will leave the land empty or even sell it to become landless. Therefore, FLA does not contribute much to their livelihoods any more. As a consequence of the available investment capacity, FLA has indifferently widened the gap between the poor and ethnic minorities with the rich people in terms of access to natural assets. In other words, FLA does not support much for the poor and ethnic minority groups in reducing their poverty circumstances.

3.2 METHODOLOGY

3.2.1 Research Site

The research site was Thuong Nhat Commune located in upland Nam Dong District of Thua Thien Hue Province in Central Vietnam. The commune is comprised of 7 villages, namely, from village number one to village number seven, having 470 households and 2026 people. Co Tu ethnic minority people, who have lived in Thuong Nhat for a long time, account for 92% of the total population living in the commune. The Kinh majority people with 8% of the population came from the lowland Phu Loc District to settle down in Village Seven of the commune when new economic zones were established after liberation day in 1975. Village Seven is also central to the commune where the CPC headquarters, schools, health care center, and some restaurants are located (Nam Dong Department of Statistic, 2010). According to older people in Thuong Nhat Commune, the Co Tu ethnic people originated from remote Hien and Giang districts of Quang Nam Province. For several different reasons, they migrated to Thua Thien Hue Province 600–500 years ago, and dwell mainly in A Luoi and Nam Dong districts. The total natural area of Thuong Nhat now is 11,377 ha, of which land for agricultural production accounts for 3.5% while forestland is over 86% (Fig. 3.1).

3.2.2 Methods

Both quantitative and qualitative studies were used for data collection of the research. The series of tools for data collection used in the research were comprised of observation, key informant with in-depth interview, group discussion, and the household survey.

FIG. 3.1   Location of the research area. *Collected from Google Map, 2011.*

3.2.2.1 Sampling

Group discussion sampling: Group discussion was the most important tool of the research and also the first activity in the field for data collection. Participants of group discussion were divided into four groups: group 1 had 15 participants from Village One and Village Two; group 2 had 16 villagers from Village Three and Village Four; group 3 was comprised of 18 farmers from Village Five, Village Six, and Village Seven; and group 4 included 14 villagers from all over the commune plus 2 staff from Thuong Nhat CPC.

In-depth interview sampling: Thirteen key informants from village to district were selected for in-depth interview. They comprised of (1) seven village leaders of seven villages in Thuong Nhat Commune; (2) one chairperson of Thuong Nhat CPC; (3) four leaders from the Nam Dong Department of Forest Protection, the Department of Natural Resources and Environment (DONRE), the Department of Agriculture and Rural Development (DARD), and the Agro-Forestry Extension Center; and (4) one vice chairperson of Nam Dong CPC. These people are leaders and policy makers at local levels who have had certain influences on socioeconomic life in their locality.

Household survey sampling: Based on the list of Thuong Nhat households including the poor and the nonpoor of 1995—the year when FLA was first implemented in Thuong Nhat Commune—provided by Thuong Nhat CPC, 95 samples (accounting for 45% of its households in 1995) were chosen randomly for household interviews, 50 of whom are poor and 45 of whom are nonpoor households.

3.2.2.2 Data Analysis

Data collected from household surveys was coded and analyzed by Excel and Social Science Statistical Program (SPSS) softwares. Descriptive statistics and compare means were used in analysis including percentage, frequency, mean, standard deviation, statistical significance of variables at the points of time before (the year 1995) and after FLA (the year 2010).

3.3 RESULT

3.3.1 Livelihood Activities

3.3.1.1 Production Activities

The livelihood activities of the poor households in Thuong Nhat Commune were listed and analyzed under three categories: on-farm activities, off-farm activities, and nonfarm employment. The results were zoomed out briefly in Table 3.1.

In the commune, on-farm activities referred to the production aspect of households, including crop cultivation such as the growing of annual food crops, yearly industrial crops, fruit trees in home gardens, perennial plantations of both industrial and forest trees, and livestock husbandry such as the raising of pig, cow, buffalo, poultry, and freshwater fish. Off-farm activities related to the implementation of natural forest protection contracts between households and BMNP, the protection of community forest, the collection of NTFPs, and the illegal exploitation of timbers and wildlife from natural forest. A minority of laborers also have been working in the off-farm sector, such as the state's officers and wage laborers. They were paid a full-time salary or daily wage based on the task and working time. Nonfarm employment mainly consisted of small-scale industries like constructors, carpenters, tailors, and small trades such as retail trades, restaurants, and other workers at home.

3.3.1.2 Distribution of Labor

During the discussion, aside from the enumeration of household livelihood's activities, participants also assessed the distribution of labor to the different main livelihood activities as well as made some remarkable points that affected significantly livelihood activities. First, there was no clear difference in labor distribution between Co Tu poor households and Co Tu nonpoor households in general. Most of them spent a lot of time performing on-farm activities such as work on the agricultural land and forestland and cultivating the home garden. The time they spent on off-farm activities accounted for only a small percentage, whereas Kinh households allocated more labor for off-farm activities than for on-farm cultivation (Table 3.2). The different distribution of labor between these two ethnic groups reflected clearly the difference in their occupations: Kinh people's work was dependent mainly on trades and services, while Co Tu household's jobs focused predominantly on crop cultivation. Second, different crops were cultivated not only for basic food demand of households but also for sale at the market. Most fruit trees and short-term industrial crops were grown on home garden land. Rice was planted on wetland as paddy rice, and on hilly land (intercropping with acacia and rubber trees) as traditional rice or dry rice. On the forestland, people

TABLE 3.1   Livelihood Activities of Households

Sector	Activities
On-farm	
• Annual crop production	Growing annual food crops: Wet rice, dry rice, cassava, maize
	Growing yearly industrial crops: Sugar cane, peanut, green bean, chilly, vegetables, zinger
• Perennial crop production	Planting fruit trees in home garden: Bananas, papaya, jack fruit, betel-nut, orange, lemon, mango, guava, pepper
	Planting perennial plantations: Acacia, rubber, bamboo for shoots
• Livestock raising on farm	Raising cattle: Pig, cow, water buffalo, goat
	Raising poultry: Chicken, duck
	Fish raising in fish pond
Off-farm	
• Collection of forest products	Exploiting nontimber forest products: Bee honey, rattan, bamboo-shoots, medicinal plants, fish
	Illegal timber and wildlife exploitation: Cutting timber down, hunting, carrying products to home for sale
• Harvest of plantation woods	Temporary procession of acacia wood for sale
• Forest protection	Natural forest protection contract with BMNP
	Community forest management
Nonfarm	
• Job at the state's organizations	Work for the state such as teacher, forest guard, nurse, communal officer with full-time monthly salary
• Wage employment	Provide labor for other households with daily wage
• Small-scale trade, industry at household level	Provide services or self-employment for other off-farm activities such as retail trade, carpentry

Source: Group discussions, 2011.

grew acacia and rubber trees. Cassava, like one of the main crops, was intercropped with plantation trees of acacia and rubber on forestland and around the home garden. Third, aside from the cultivation on their land, a majority of household laborers spent more time accessing the natural forest for additional income, although some of them did not perceive that these forests belong to the community or BMNP management while others knew these were illegal activities. After FLA, the people's right to use natural forest resources have changed; indeed, as they were no longer allowed to practice slash and burn production, collect the timbers, or hunt wildlife anymore, which they had done frequently before 1995, FLA did not yet apply

TABLE 3.2   Labor Distribution by Ethnic Groups

Activities	Co Tuhhs (%)	Kinh hhs (%)
On-farm cultivation	50	25
− Annual food crop	40	
− Home garden	15	
− Rubber plantation	20	25
− Acacia plantation	15	25
− Livestock husbandry	10	50
Subtotal	100	100
Off-farm practice	30	25
− NTFPs collection	50	50
− Illegal timber exploitation	40	50
− Forest protection contract implementation	10	
Subtotal	100	100
Nonfarm employment	20	50
− Hired labor	80	20
− Small-scale trades and services	20	80
Subtotal	100	100
Total	100	100

Source: Group discussions and case study, 2011. (During group discussion of group 3 (with people who came from village number five, village number six, and village number seven), one Co Tu and another Kinh member were chosen to calculate the labor distribution of their households in a year. Members estimated the percentage of their household labor force that was spent for each production activity through scoring it. Total scores are 10. Each score is equivalent to 10% of the labor force for activity. The results were fulfilled by other participants when presented, and became the final product of group discussion.)

in the commune. Fourth, many different crops were found to be unsuitable on the land and environment around it, particularly fruit trees on home garden land and rubber trees on steep forestland near the natural forest. Of the many reasons mentioned, the bad quality of the soil and harmful pest attacks were included. Income from these sources was quite limited and this only contributed partially to the food or cash shortage of households. People therefore had to spend more time to search for other income sources in other employment sectors instead. Finally, the distance to the field, particularly to plantation forest area, was quite far away from the home. In several cases, it took about one to one and a half hours on average to walk to the field for work. The daily working time therefore was reduced due to the long distance to home from the workplaces. In fact, the local people have little chance to receive land close to their residence area.

Livelihood outcomes of poor households lie partially on their income sources, including the number of income sources that households have and the distribution of income sources in the income structure of households at the time of the research.

3.3.1.3 Number of Income Sources

Household surveys indicated that households in Thuong Nhat Commune had about 14 sources of income (Table 3.3). There was no significant difference in the number of sources of income between poor and nonpoor groups because most of them had all types of income sources, except poor households have not been taking part in the small-scale trades and services sector like nonpoor households. Based on an analysis of the proportion of households having income sources, it could be seen that home gardens, which are second source after cassava, were the most common income sources with over 95% of poor households and 100% of nonpoor households having them; following were paddy rice and livestock with over 80% of households in both wealth groups having these two sources. NTFPs collection and hired labor provision were also popular income sources with over 70% of both wealth households having these sources. Growing of perennial plantation with acacia and rubber trees seemed to be one of the new income sources after FLA when the state not only encouraged local people but also supported them with cash and materials as well as technical assistance simultaneously to plant these two species on allocated forestland. Over 30–50% of poor households had income sources from growing rubber and acacia plantations, while this percentage in nonpoor households was over 55–70%, respectively. Timber exploitation, together with NTFPs collection, was the traditional income source of upland people whether this was legal or not. However, the exploitation of natural forest requires households to spend a lot of time and labor; therefore, it does not attract many people to enjoy it while a couple of households could

TABLE 3.3 Different Income Sources of Interviewed Households

Income Sources	Percent of Households (%)	
	Poor (*n* = 50)	Nonpoor (*n* = 45)
1. Home garden	96	100
2. Paddy rice	100	87
3. Cassava	100	100
4. Livestock	82	87
5. Fish pond	10	13
6. Acacia wood	52	73
7. Rubber latex	32	56
8. Forest protection	14	27
9. Timber exploitation	36	40
10. NTFPs collection	90	73
11. State work salary	30	18
12. Retired salary	26	16
13. Hired labor	96	73
14. Small trades, services	0	22

Source: Household surveys, 2011.

not address these requirements. In fact, only 35% of poor households and 40% of nonpoor households could access this income resource. Also, although salaries from the state for full-time working officers or retired people are the most stable sources of income and contribute a high proportion of the income structure of the commune, there were only a small number of both wealth households having income from these sources, with about 26% of poor households and around 17% of nonpoor households. Other two productive activities that contribute to the total income of households in Thuong Nhat Commune were fish raising and natural forest protection contract implementation with BMNP, although the percentage of households participating in these two activities were not too high, with only just over 10% of both wealth households raising fish in fish ponds as well as over 14% of them implementing natural forest protection contracts. As the above research has somewhat analyzed, there were few households in the nonpoor group having income sources from small-scale trades and services with the percentage being 22%. In spite of evidence about the small number of households taking part in the small-scale trades and services sector, results from group discussions showed that this was one of the substantial nonfarm activities of almost all Kinh households, and also contributed a large proportion of the total income of Thuong Nhat Commune.

3.3.1.4 Income Distribution

In terms of the contribution of income sources for the economic development of Thuong Nhat households (Table 3.4), the analyzed results of household surveys showed that acacia wood and rubber latex from plantations were two types of product that generated the highest cash income, with 29% and 37% of the total income in poor households and the non-poor group, respectively. Fruit tree and annual crop products were the second highest income within both wealth groups where they contributed equally 26% of the total income of poor households and 25% of the total income of nonpoor households. Among fruit tree and annual crops, cassava was the main product, accounting for over 15% of the total income of the poor households and 17% of the total income for the nonpoor households. Second was the paddy rice product, whose share of total income for the poor group was around 7% and for the nonpoor group was 6%. Third were home garden products, which contributed only around 2% of the poor's total income and nearly 4% of the nonpoor's total income. These results were related to the observations that cassava was intercropped everywhere on the forestland together with acacia and rubber trees, the size of the paddy land for rice owned by households was small, and fruit tree products from around home gardens depended mainly on bananas and betel nuts that can always be attacked severely by harmful pests.

The same paddy rice crop and livestock product including cattle, poultry, and fish were also somewhat important income sources for all wealth groups, although they contributed far less to the household total income; around 7% for the poor households and 9% for the nonpoor group. Off-farm products such as timber and NTFPs collected from natural forest, and cash from natural forest protection contract implementation played an important role in the household economy although the percentage of households taking part in these activities was not too high. Products from natural forest protection and exploitation contributed 10% of the total income of poor households and nearly 9% for the nonpoor.

Finally, the contribution of nonfarm activities to total income, including daily hired labor, full-time salary officer, and small-scale trade services, was more important in the poor group than the nonpoor households. It was around 26% for the poor and only 18% for the nonpoor.

TABLE 3.4 Income Distribution at the Household Level

Income Value	Poor (n = 50)	Nonpoor (n = 45)
	Percent (%)	
Fruit tree and annual crop production	26.8	25.8
1. Home garden	3.8	2.1
2. Paddy rice	7.1	6.1
3. Cassava	15.9	17.5
Livestock husbandry	7.0	9.5
4. Livestock	6.8	9.4
5. Fish pond	0.2	1.0
Plantation crop production	29.4	37.7
6. Acacia plantation	16.5	21.5
7. Rubber plantation	12.9	16.3
Off-farm production	10.1	8.8
8. Forest protection	1.1	1.1
9. Timber exploitation	2.5	4.3
10. NTFPs collection	6.4	3.4
Nonfarm production	26.8	18.2
11. State work salary	12.8	4.3
12. Retired salary	9.0	6.3
13. Hired labor	5.0	3.0
14. Small trades, services	0.0	3.5

Source: Household surveys, 2011.

As explained above, although only a small number of households have salary from the state, it contributed a larger proportion to the total income for them with around 23% for poor households and nearly 11% for the nonpoor group. Among off-farm income sources, wage labor contributed a very small percentage to the total income of households, with 5% for the poor and only 3% for the nonpoor. Of course, the small-scale trades and services sector contributed nothing to the total income of poor households not one of them took part in this activity, whereas it contributed nearly 4% of total income of the nonpoor households.

3.3.2 Current and Potential Risks in Livelihoods of the Poor Households

Data and information about the risks to the livelihoods of poor households were gathered from in-depth interviews and group discussions. Referring to the risks would seem to be evidence showing that local people paid much attention about their life related closely to FLA consequences currently and tin he future.

First, it became more clear during the group discussions that the number of cattle such as cows, water buffalo, and goats kept by households decreased enormously over the last 10 years. The reason was due to the establishment of a local regulation to punish farmers if their cattle destroy the young plantation forests, and open access to pasture for keeping cattle was no longer possible as before. For most of villagers, the reduction of domestic animals persuaded them to focus more on their forestland instead of on livestock husbandry on one hand, but the limitation of open access to pasture for cattle raisings also increased the vulnerabilities for those who have no forestland but depend on livestock husbandry for improvement of their livelihoods on the other hand.

Second, it appears that the income of a large number of people now depends mainly on forest plantations because many agricultural land areas were transformed into forestland types such as land for planting local cassava or dry rice crops became forestland for planting industrial cassava intercropping with acacia and rubber trees. Furthermore, farmers now put many resources to work for forest plantations such as labor, time, and material inputs (seedlings, fertilizers). The transformation of land-use purpose can lead to a severe shortage of food not only for human consumption but also for domestic animals. In additon, the traiditional indigenous varieties of agricultural crops are likely to be lost or threatened due to having no suitable habitat in which to grow. The high costs of investment for forest plantations can push the poor toward jumping into the debt-driven situation. Thus, as a result, a potential risk will appear for plantation forest-dependent people life if negative fluctuations of market prices of plantation products occurs, or pests happen to attack their plantations, or even typhoons appear to destroy their forests. These will likely lead them to a crisis of income that will not be easy to overcome in the long term.

Lastly, unappropriate FLA planning of local authorities was implemented, as several villagers mentioned, and has been causing the phenomena of "forestlandlessness" for a couple of young households in Thuong Nhat Commune. According to Thuong Nhat CPC, most land in the commune includes agricultural land and forestland now allocated to households already, and it has no more to allocate to new households. Having not enough land for cultivation is one of the biggest difficulties that not only young people but also all in the Thuong Nhat community have to face because serious illegal land encroachment for cultivation and exploiting natural forest for additional income by this class of landless people has been occuring there. This will likely lead to a crisis in production land for local farmers and authorities in this upland area in the future.

3.4 DISCUSSION

From the research, there are a number of recommendations to make, including

1. Check and review all existing types of land of the commune to have both long-term and yearly plans of land use that are suitable with the trend of socioeconomic development of the upland region in general and of the commune in particular in the coming years. Approaches of advocacy and lobby should be applied to the BMNP and the Nam Dong Protection Forest Management Board (PFMB) at the same time with the hope to have more forestland for local cultivation.
2. Support local villagers to find suitable tree species for forest plantation. On one hand, establish the niche market for biodiversity products from annual crops and perennial

trees, and promote the commercialization and domestication of indigenous species in agroforestry on the other hand.

3. Strengthen the capacity of local people in the field of development linkages with NRM to have suitable intervention. Further interventions like markets, credits, incentives, and production services should be discussed with villagers to promote better cultivation and crop systems on the forestland.

4. Networking, capacity building, and learning are essential for the short, middle, and long-term processes of sustainable household livelihoods (Mittelman, 1997). Therefore, by fostering networks among local people, community-based organizations (CBOs), government organizations, and nongovernmental organizations, those working in the commune should be promoted to provide more opportunities for improvement of local people's abilities. This will contribute an important part to the consolidation and enhancement of local livelihoods, especially livelihoods of the poor households.

However, the research also provides a general discussion on the equity in FLA, the influence of FLA implementation to different household groups, and the sustainability of FLA in Thuong Nhat Commune and Nam Dong District. The following discussion will contribute an important part for the research findings to more clearly answer the research questions.

3.4.1 Equity or Inequity in FLA?

Some previous research concluded that poor people always gained least from FLA, including both quantity and quality of forestland, compared with those who are better off or have stronger power and higher positions in society. This leads to the unfair competition in allocation of forestland among different wealth status groups. In their research, Phuong (2004) and Sunderlin et al. (2005) emphasized that poor households did not receive more forestland for afforestation. Therefore, FLA appears not to have assisted the poor households to overcome their poverty and to have widened the gap between the poor and the rich. However, the findings of this research showed that FLA in Thuong Nhat Commune was open access for all households who wanted to take part in reforestation programs funded by the government as well as for those who could invest to develop the plantations by themselves. As mentioned earlier, the poor groups did not pay much attention to FLA until they recognized the value of forestland later due to the fact that it offered the possibility of long-term benefits on one hand and was compounded by low market prices and limited production investment capacities on the another hand. Therefore, they no longer received the land. As a consequence, state officers and better-off households were the first people who received forestland:

> We spent much time, we organized several campaigns to propagandize the FLA to villagers and persuade them in receiving the land and the forest for plantations, but at last they did not. They do not want to be allocated the forestland at all. For this reason, in order to implement successfully the FLA program, we have to become the pioneers in allocation of the forestland. *In-depth interview, 2011—the statement of an officer in Nam Dong District who now owned 30 ha of forestland.*

Correspondingly, aside from having a social relationship, it could be said that FLA to households depends totally on the production capacity of each household. In other words, FLA was reserved for households who not only have the financial means for investment but also have social networks to access freely the capital supply from the government.

Regarding the size of each forestland plot and the number of plots owned by each household, research findings indicated that many small pieces of forestland were allocated to households and many households possessed more than one plot. In the case of facing difficulties if natural disasters occur, or even not having enough capital to invest for production, it is very easy for poor households to sell some their plots of land and become landless. It can be predicted that this process will be exacerbated more if the better-off households attempt to accumulate the forestland by buying from the poor and become the people possessing the most forestland. In this regard, a question can be posed: What policies for poor households need to be established to mitigate this potential risk? Local authorities and policy makers should carefully consider this critical matter to fill the gap.

3.4.2 Old Households and Young Households, Who Were Affected the Most by FLA?

> Before the year 2000, my family did not pay much attention about FLA because if I have received forestland, I would not have had any sources to invest for it. In addition, plantations needed a long term to give the benefits. However, my neighbors earned very high income from investing in their acacia and rubber plantations recently. In late 2006, my family was informed again to participate in FLA of the commune which was funded by WB3 project, and now I have 1.4 ha. I also easily got a loan from the Bank of Agriculture and Rural Development (BARD), which pushed me to invest more in acacia plantation. Of course, I had to use my Red Book as collateral to apply for this capital. *Statement of a farmer in household surveys, 2011.*

According to data recorded at the Thuong Nhat CPC office, the poor household rate of the commune in 1995 was 67%. Up to 2010, this percentage was decreased to 20%, decreasing 47% compared with the time when FLA was implemented for the first time there. Obviously, this comparison has relative significance only due to the difference in poverty criteria at different points in time. Although there is no quantitative data to prove it, the CPC asserted that FLA has played an important role in the reduction of the poverty rate in its locality. In this research, it found that after 15 years of FLA implementation, under its positive influence, the poverty rate in poor households reduced sharply to 86% (Chart 3.1). A large number of poor households have now escaped successfully out of poverty since they took part in the FLA

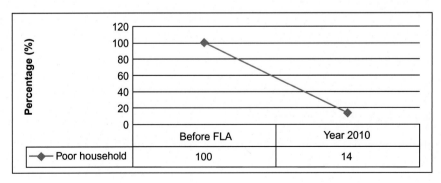

CHART 3.1   Change in poor household percentage before and after FLA (*n* = 50).

programs. It could be said that FLA implementation has partially filled the gap between the poor and the nonpoor groups in the research commune.

My research findings are contrary to the previous studies of Tuan (1999), Phuong (2004), or Sunderlin et al. (2005) when these authors concluded that Kinh inhabitants and other influential powerful ethnic groups have had a tendency to get more benefits from FLA in contrast with the poor and ethnic minorities, and FLA does not offer them much support in reducing their poverty condition. They explained that when poor and ethnic minority households do not have enough financial sources to invest for afforestation, they will leave the land empty or even sell it to become landless. Therefore, FLA no longer contributes much to their livelihoods. The difference in research findings between researchers could be explained by the credit supports from the state banks for the poor households as they have easier access to several credit sources to get the loan for their production investment on the allocated forestland and forest plantation. The statement of the head of village number four, "If to have clear production development plans, and if they are approved by the CPC, our poor villagers can submit their loan document to the BARD or BSP and easily get money from them to plant the forests", is evidence confirming that poor households in Thuong Nhat now do not worry much about the capital to invest for their production.

Another issue of FLA influence to the poor households that needs to be discussed is the current landless phenomena of newly separated young couples. As explained above, the Land Law that regulates land-use rights of households expires after 50 years with forestland, and 20 years for agricultural land, so households are fully responsible for their own land, and no people other than their household members can harm or cultivate it. Therefore, local authority could not adjust FLA by revoking the land of one person to reallocate to another household. It is not uncommon to see that several young couples in Thuong Nhat have no land for cultivation; some of them even have no piece of land for housing and have to borrow the land of their relatives to live on. This scenario will be worse under the pressure of incessant population growth at present. As for the poor farmers in the upland area, having no types of land to cultivate, what will they do to survive? In this regard, a big question that arises for policy makers at any level is what will they support or what regulations will be established to assure that new poor households can live without land in the near future.

> Lack of land for cultivation is one the biggest constraints of our commune. Agricultural land is limited while most of forestland belongs to BMNP and Nam Dong PFMB. Therefore, we did not have enough land any more to allocate for young couples who just separated from their parents. *Statement of Thuong Nhat CPC vice chairwoman in in-depth interviews, 2011.*

In terms of new landless households, the research found that a couple of households in Thuong Nhat Commune still have reclaimed the forestland of Nam Dong PFMB, which belonged to their forefathers in the past. This land encroachment is illegal activity following the Law of Forest Protection and Development, but local people have continuously cultivated it first and will hand it over to their young children later with the expectation that, once they encroached upon the land, the state will allocate it for their younger generation to gain a living. Thus, it could be assumed that for the near future illegal land encroachment in the state forest areas will continue and the landless phenomena in young families will continue, and serious conflicts over forestland possession between poor people and the state will arise that need to be solved. Who will be affected the most and who will lose the most in this case? A further survey is required to inform FLA policies on a suitable adjustment for the issue.

3.4.3 FLA-Based Livelihoods of the Poor: Sustainability or Not?

The livelihoods of the poor households in Thuong Nhat commune have already been transformed, and this section discusses some of the effects of FLA implementation on the lives of local people as expressed in three aspects of sustainable development: the environment, society, and economics.

3.4.3.1 *Issue of Environment*

One FLA objective is the regeneration of degraded forest barren land and barren hill recovery and the likelihood of achieving this in Nam Dong District. In this regard, according to Nam Dong DONRE (2010), up to the end of 2009 the forest recovery in Nam Dong District reached 79.7%, quite a high proportion considering the forest cover of the northcentral region of Vietnam was just 38% at the same time. The forest cover in the research area is being increasing quickly, but at what levels of diversity and quality?

At a grassroots level, the group discussions and in-depth interviews revealed that most of the forest plantations on forestland in Thuong Nhat Commune, which lie on the hillsides, were more and more poorly developed with exotic and monoculture species like acacia and rubber trees. Barren land and barren hills were planted as plantations but with poor quality and low biodiversity. Although local people suggested that this area is known to be the center of typhoons and the origination point of big floods, it needs to protect the environment to reduce these disasters, as noted by the statement of a woman in her interview: "From the year 2002 up to now, several floods and storms have destroyed many houses, home gardens and broke thousands of rubber and acacia trees on the plantations in Nam Dong district. Therefore, it is quite important to plant and protect the forest because it will help to decrease the natural disasters" (Household surveys, 2011). In fact, local people still continuously invested their capital in forest plantations with alien species because of the high-income return. For this reason, it could be said that the movement from livelihoods focused on self-consumption production to livelihoods focused on market-driven perennial plantations can bring cash income for poor households and help them to escape out of poverty on one hand, but it can be a factor of potential risks and challenges for the forest functions of protection with the growth of nonnative plantations on the other hand.

Another point associated with the conservation issue that needs to be discussed is the potential danger in the loss of native genetic sources. Several older people mentioned that although giving low and uncertain yields, a large number of their native varieties of agricultural crops such as Luamua (seasonal rice or dry rice), Ba Trang cassava, and local maize always have been respected and protected by several Co Tu generations. These are local varieties that not only have fed Co Tu people from generation to generation but also have been a precious cultural symbol of the people in the past and at present. However, most of these varieties now are standing at the bank of extinction. What is the reason?

As explained earlier, the program of sedentary cultivation forced people to cultivate in the same plots of land from year to year. The soil therefore has no time to regenerate and became unfertile. As a consequence, native species of annual crops could not grow up well on the type of unwoody land, and the low yields were continuously increased step by step. For this reason, poor farmers were not encouraged to use these native species for food security any more. Whereupon they started to use other varieties freely supported by the state that have high yields and short harvest cycles instead. In addition, many poor households now have converted their agricultural land to forest plantations because the soil of agricultural fields is getting poorer on one hand and income from forest plantations is higher on the other hand. Agricultural land was transformed into land for perennial plantations that not only produced less food for local people, but also provided no room and space for conservation of the native crop species. In fact, this issue is a signal that conflict is occurring between ecological conservation and economic development in Vietnamese upland regions at present. As Scherr and Dewees (1996) suggested, the complex relations between conservation and development, between natural resource base and livelihoods, and the interactions among household production, household welfare, and markets and the environment need to be considered carefully. Therefore, this issue needs to be studied more to find the appropriate solution for the trade-offs between them.

3.4.3.2 Issue of Society

FLA implementation was recognized for bringing many positive benefits to the social life of inhabitants in Thuong Nhat Commune in general, as it was highly perceived and appreciated by local people for the increase of household income and living standards, the better access to schools of children, the improvement of household access to common services, the free supply of information, and the higher position and voice of women in the household and in the community.

As many villagers mentioned, the close collaboration among local households and state organizations, CBOs working in the villages, and the commune has showed good improvement in social network participation of household members, especially for female members in the networks. These have established and developed through a lot of capacity building efforts of many stakeholders and institutions in the research area. The positive change in knowledge, skill, and experience of local people promises to bring a successful enhancement of livelihood sustainability to them at the grassroots level.

However, in an opposite aspect, the FLA program also brought a negative effect to the Thuong Nhat community as a number of young families fell into a bad situation of landlessness and unemployment as explained by a young landless household head when he mentioned about his family jobs: "If you are landless person here, you have three options to choose: i) stay continuously with your parent and use their land, ii) go to the natural forest to collect NTFPs, or to work for some households as hired labors for daily income, and iii) go to the big cities in the South such as Da Nang or Ho Chi Minh to find a job there. In fact, we still do not know what we will to do to survive in the coming future" (Household surveys, 2011). As discussed above, this problem will be exacerbated more if no suitable solution is found to solve it while the pressure of population growth will be even higher in the near future. Thus, an assessment about the negative effects of FLA on the social life of young families here is necessary for further research.

3.4.3.3 Issue of Economy

My house was built by money from the selling of rubber latex, my motorcycle [sic] was also bought by money from the sale of my acacia plantation. If I have more forestland, I will plant rubber plantation continuously. If I have more capital, I will also invest it in acacia plantation continuously. In brief, acacia and rubber plantations are the most important income sources for farmers like me in my area at the moment. *Statement of a household head in household surveys, 2011.*

As local villagers admitted, the high income from forest plantations has allowed them to escape out of poverty and reduce illegal forest exploitation. In the livelihoods of households, income from home gardens, livestock husbandry, paddy rice crops, and NTFPs helped them to make a daily living while income from perennial plantations with rubber and acacia trees supported them further to have savings and long-term benefits. Forest plantations are seen as providing a promising prospect for high income, contributing to a sustainable economy in the commune in the coming years.

However, as mentioned before, the market price of forest plantation products was one of the most important factors affecting the income of local people in Thuong Nhat Commune as there have been more and more households dependent on it. Although local households who have been engaged in forest plantation have benefited through increasing income, improving livelihoods due to high prices of forest plantation products, but the research findings suggested that some risks and vulnerabilities are implicit in their economic life. First, the present high prices of plantation products (here are acacia woodchips and rubber latex) have led the Thua Thien Hue provincial authority to expand its plantation forest areas to around 1490 ha from 2007 to 2012 (TT Hue DARD, 2007). It can be expected if this program has been implemented successfully, that the abundance of acacia woodchips and rubber latex in the next few years will lead to a decrease in the market price of these products. Although the market prices have increased in the last 5–7 years, no future forecasts in supply and demand of plantation products have been undertaken. A potential risk arises for plantation owners if negative fluctuation of market prices occurs for these products. Lessons learned from the cases of sugar cane and cashew programs in Vietnam in the period 2000–05, and their impacts on sugar cane growers and cashew planters suggest a similar scenario for forest plantation products due to oversupply (Tam, 2008). Second, a constraint connected to monoculture of plantation species is its vulnerability to noxious pests. The eucalyptus plantations in some places in Nam Dong District had already experienced the negative effects of serious pests in the past. In addition, serious attacks of harmful pests to plantations of acacia have already been reported in Quang Tri, a province nearby, since 2005, and it was already widespread to Xuan Loc, a commune not far from the research area, which was evidence of making forest growers more vulnerable due to the damage of plantations (In-depth interviews, 2011). Therefore, such current and potential risks as well as vulnerabilities of forest plantation products need to be further researched to provide appropriate advice to policy makers and for forest plantation growers in the upland areas.

In general, forest plantations have been providing opportunities to poor people for a high-income source. This prospect started to be realized in Thuong Nhat Commune as well as at several places in Thua Thien Hue Province since FLA was implemented a couple of years ago. However, selection of appropriate species of trees for planting, assurance of land tenure, and establishment of markets for plantation products are always crucial considerations for the success of agroforestry programs (Leisher and Peters, 2004).

References

ADB, 2000. Study on the Policy and Institutional Framework for Forest Resources Management. Asian Development Bank. TA No. 3255 – VIE.

ADB, 2001. Poverty Alleviation in Credit, Forestry and Sedentary Programs in Vietnam. Asian Development Bank, Hanoi.

Apel, U., 2000. Forest Protection Regulations as a Precondition for Natural Regeneration in the Song Da Watershed, Northwest Vietnam. Paper at the Workshop "Sustainable Rural Development in the Southeast Asian Mountainous Region" organized by European Committee (EC), SIDA and GTZ.

Ba, H.T., 2002. People, Land and Resources Study in Vietnam. WWF Indochina Program.

Bellamy, R., 2000. Assessing Different Approaches to Forest Management in Vietnam. International Development Research Centre, Ottawa.

Castella, J.C., Christophe, J., Boissau, S., Thanh, N.H., Novosad, P., 2004. Impact of forestland allocation on land use in a mountainous province of Vietnam. Land Use Policy 23 (2006), 147–160.

Cuc, L.T., 2002. The Issues of Development in the Mountainous Areas of Vietnam (Lecture Notes). Centre for Resources and Environment Study, Hanoi.

Due, P.N., 1999. Evaluation of the Participatory Land use Planning and Land Allocation Methodology at Communal Level Developed by SFDP Song Da. SFDP Song Da Project, Vietnam.

Dung, D., 2006. Land and Forest Assignation: Results and Solutions need to be completed. Newspaper of Vietnamese Communist Party.

Hoang, D.H., Son, D.K., 2008. Forestland and Forest Allocation in Vietnam: From Policies to Practice. Workshop Document.

Lan, L.V., Ziegler, S., Grever, T., 2002. Utilization of Forest Products and Environmental Services in Bach Ma National Park, Vietnam. http://www.mekong-protected-areas.org/vietnam/docs/bach_ma_forest_products.pdf.

Leisher, C., Peters, J., 2004. Direct Benefits to Poor People from Biodiversity Conservation. The Nature Conservancy, Brisbane.

MARD, 2008. Forest Sector Manual. Ministry of Agriculture and Rural Development. Retrieved May 20, 2010, from http://www.vietnamforestry.org.vn/.

Mittelman, A., 1997. Agro and Community Forestry in Vietnam: Recommendations for Development Support. Report for the Forest and Biodiversity Program, Royal Netherlands Embassy, Hanoi.

Nam Dong DONRE, 2010. Report on Community Based Natural Forest Management in Nam Dong District. Workshop document.

Nam Dong Department of Statistic, 2010. Statistics in 2010. Statistical Yearbook 2010 of Nam Dong District. Thua Thien Hue Province.

Ngan, L.T., Tho, N.T., 2000. Report on Poverty Situation of Tay Ethnic Communities in Bac Giang province. Hanoi Agricultural University, Hanoi.

Nhan, T., 1998. Forestry: A Way to Improve the Lives of Mountain People. Vietnamese Forestry Magazine. September edition.

Phuong, P.X., 2004. Consultancy Report on Survey, Assessment on the Implementation of Benefit Sharing Policy for Households, Individuals and Communities Allocated, Contracted Forests and Forestland in Gia Lai, Dac Lac, Son La, Dien Bien provinces. Ministry of Agriculture and Rural Development, Community Forestry National Working Group, Hanoi.

Scherr, S.J., Dewees, P.A., 1996. Policies and Markets for Non-timber Tree Products. EPTD Discussion Paper No. 16. IFPRI, Washington DC, USA.

Sikor, T., 2000. The allocation of forestry land in Vietnam: did it cause the expansion of forests in the northwest? Forest Policy Econ. 2 (1), 1–11.

Sunderlin, W.D., Angelsen, A., Belcher, B., Burgers, P., Nasi, R., Santoso, L., Wunder, S., 2005. Livelihoods, Forests, and Conservation in Developing Countries: An Overview. World Dev. 33 (9), 1383–1402.

Sunderlin, W.D., Ba, H.T., 2005. Poverty Alleviation and Forests in Vietnam. Center for International Forestry Research, Jakarta.

Tam, L.V., 2008. The Forest Land Use and Local Livelihood. Case Study in Loc Hoa Commune, PhuLoc District, Thua Thien Hue Province. Master thesis in rural development with specialization in livelihoods and natural resources management, Hue University of Agriculture and Forestry.

Thanh, N.X., 2000. Report on Issues Concerning Forestland Allocation to the San Diu Ethnic Minority Farmers in Bac Thai Province. Hanoi Agriculture University, Hanoi.

Thanh, T.N., 2001. Participation in Land Use Zoning, Forestland Allocation and Management in DakPho. Daklak. MRC/GTZ Sustainable Management of Resources in the Lower Mekong Basin Project.

Thu, N.T., 1999. Report on Socioeconomic Situation of the Tay Ethnic Minority in Bac Giang Province. Hanoi Agricultural University, Hanoi.

Thuan, D.D., 2005. Forestry, Poverty Reduction and Rural Livelihood in Vietnam. Retrieved May 20, 2011, from http://socialforestry.org.vn/Document/DocumentEn/Forestry%20Poverty.pdf.

II. LAND USE SYSTEM AND LIVELIHOOD COMPLEXITIES

TT Hue DARD, 2007. Detailed Planning of Commercial Forest Plantation Program 2007–2010. Project Document. Hue: Department of Agriculture and Rural Development.

Tuan, P.D., 1999. A Comparative Study of the Socio-economic Situations of a Lowland Commune and an Upland Commune in the Northern Mountains. Agricultural University, Hanoi.

Tuyen, V.D., 2010. Real Situation and Solutions for Forestland and Forest Allocation in Vietnam. Workshop Document, Ministry of Agricultural and Rural Development, Hanoi.

Vien, T.D., Huong, P.T., Rasmussen, Schultz, M., 2007. The Social and Environmental Dimensions of Changes in Land Use in the Ca River Basin, Vietnam. In Institutions, Livelihoods and the Environment: Changes and Response in Mainland Southeast Asia. Compiled by Straub, A. Copenhagen: Nordic Institute of Asian Studies.

Changing Land Access of Resettled People Due to Dam Construction in Binh Thanh Commune

P.T. Nhung, T.N. Thang

Hue University of Agriculture and Forestry, Hue, Vietnam

4.1 INTRODUCTION

Dams play an important role in energy production, flood control, and agricultural drainage. By the end of the 20th century, there were over 45,000 large dams in the world, built across 140 countries. Between 40 and 80 million people have been displaced by dam development (WCD, 2000).

There are currently around 800 dams in Vietnam (Tuyet, 2011). Dam development is a complex process that has significant impacts on people residing in the area where the dam is built (Ellis, 2004). Between 2002 and 2006, some 21,580 Vietnamese households (HHs) were resettled as a result of dam development (Ly, 2009).

In the dam development context of Vietnam, ThuaThien Hue Province has built 14 small and medium dams to support electric energy and two big dams to control floods and dry season flows. Some 1000 HHs have been resettled. In 2004, 747 HHs were resettled as a result of the construction of the Ta Trach Dam, and 51 HHs were resettled in 2005 as a result of the construction of the Bien Dien Hydropower Plant.

Natural resources are fundamental assets in rural livelihoods. Access to them, therefore, clearly has an impact on livelihood strategies and diversity (Ellis, 2004). Natural resource access (NRA) can be understood from four perspectives: quantitative, qualitative, temporal, and legal. The quantitative perspective refers to the diversity and size of resources that affected people have access to, while the qualitative aspect refers to the quality of resources and efficiency of their use. The temporal aspect refers to the amount of time people can use natural resources, and the legal perspective reveals the rights under which people can (or cannot) exploit resources. NRA can provide people with an exit route from poverty (Ellis, 2004). This means that to build and develop a sustainable livelihood, the NRA of HHs or communes needs to be identified, especially within resettlement areas (RAs) where NRA has changed.

The lives of the resettled continue to be difficult and risky. There are some programs to support and enhance livelihoods of affected people, but these have been largely unsuccessful. This study focuses on livelihoods based on access to agricultural and forestry resources. The resettlement and resource rights policies employed in their resettlement have not considered NRA as a constituent part of their livelihood strategies, focusing instead on cash compensation and physical asset replacement. Compensation for loss of access to land and forests is often limited and delayed in the implementation process (Ly, 2009). Hence, the NRA of affected communities is closely dependent on these policies and schemes, which may impede NRA. Studying how this happens is both important and necessary, so that resettlement planning and implementation can be strengthened and improved.

4.1.1 Dam Development

From a Vietnamese perspective, dams are becoming an important means to meet the needs of a modern society (Nga, 2010). Dam development aims to provide electricity and to supply water for agricultural activities and industrial and urban consumption, as well as create strategic investments that deliver multiple social and economic benefits for the country (Yen, 2003). Dam development, however, transforms natural habitats and displaces human, flora, and fauna communities (WCD, 2000). At present, Vietnam's hydropower sector constitutes a large portion of the country's electricity production. According to the Swedish International Development Agency (SIDA), hydropower contributes 75% of total electricity production. Hydropower development (especially small hydropower) is identified as a priority in the country's electricity sector strategy for 2006–2015; this strategy shows no signs of being changed for 2016–2025. By 2020, it is estimated that hydropower will generate 13,000 to 15,000 MW for Vietnam. Though in the long term, the Institute of Energy estimates that the percentage contribution of energy from hydropower will decrease markedly in comparison with other energy sources. By 2025, power generation from hydropower will contribute an estimated 14.1% to national electricity production (compared with 30.8% in 2005) (Pannture, 2008).

4.1.2 Resettlement and Compensation

Involuntary displacement is a process in which people are forced to move from one place to another due to natural disasters, environmental degradation, conflicts, or development projects (WB, 2011). Resettlement is a process to assist displaced people to replace their housing, land, and culturally or spiritually important assets; compensation can serve to give them access to resources and services and to restore their livelihoods and wealth-holding assets (WB, 2011). Involuntary displacement and resettlement are complex processes that affect not only the displaced group, but the host communities where they are relocated as well. Involuntary displacement impacts all aspects of the economy, society, environment, and health of resettled host communities (Nga, 2010).

Dam construction has been a major cause of involuntary resettlement (Tortajada, 2001). Resettlement for hydropower projects produces impacts different from other development projects. Frequently, people resettled as a consequence of hydropower development are ethnic minorities, who are poorly educated; whose lives are heavily dependent on natural resources; who have limited capacity; and who have difficulties adapting to their new environments, whether culturally or with respect to the available natural resources. These issues problematize and complicate resettlement. In Vietnam, between 85% and 99% of resettled people are ethnic minorities (Ly, 2009).

Decree No. 22/1998/ND-CP has replaced Decree No. 90/CP 1994, which defines resettlement-related issues for large infrastructure projects. The land that is compensated is the land that belongs to the area of these projects. The principle of "land for land" is applied. The resettlement agreement should be compatible with urban and rural planning and criteria for construction in these areas. Infrastructure such as roads, electricity systems, schools, and health clinics should be made available at the relocation site before resettled HHs and individuals are transferred. Sections of the decree that are relevant for people affected by hydropower projects follow.

Cultivated land lost should be compensated by an equal or bigger amount of land for cultivation in the receiving area. On the contrary, if Project Affected Persons (PAPs) receive less land than they had access to previously and/or land of lower quality, the balance shall be compensated in cash, at current rates. In addition, each HH should receive between 400 and $1000\,m^2$ of land for their house and domestic garden, with additional cash compensation if the HH receives less and/or lower quality land than they had previously. All legal documentation regarding the transfer of land rights should be handled and passed on to PAPs without any charge. All HHs should be provided with drinking water in suitable amounts and quality. Road building that supports regional socioeconomic development, including resettlement, is considered a priority investment. A road linking the resettled community to the nearest main road is defined and built. Affected people have the right to access electricity. Where the national electricity grid passes in the vicinity of the resettlement, connecting the community to it should occur quickly. If the resettlement is too far from the national grid, an alternative electricity supply (small hydro, for example) should be made available. The provision of health care and education to the resettlement is prioritized.

4.1.3 Livelihoods, Access to Natural Resources, and Policy

A livelihood comprises the capabilities, assets (including both material and social resources), and activities required for a means of living. An individual's livelihood is considered sustainable when it can cope with and recover from stresses and shocks, and when the individual can maintain or enhance his or her capabilities and assets, without undermining the natural resource base upon which the individual's livelihood depends (Chambers and Conway, 1991).

Clearly, natural capital is very important to those who derive all or part of their livelihoods from resource-based activities (farming, fishing, nontimber forest products, mineral extraction, etc.). Its importance goes way beyond this, however. None of us would survive without the help of key environmental services and food produced from natural capital, such as good quality air, clean water, and so on. Humanity depends on such natural resources, the links between them, and the continued functioning of complex ecosystems (which are often undervalued until the adverse effects of disturbing them become apparent). Degrees of access to natural resources can be directly equated with degrees of vulnerability (Ellis, 2004).

Policies and institutions operate at all levels, and in both public and private spheres, where they influence the formation and outcomes of livelihood strategies. Institutions may influence livelihoods in many ways:

• The access that poor people have to assets, the benefits they derive from them, as well as incentives for the development of assets, depend upon institutional arrangements. These in turn depend upon the institutional environment, information flows, asset characteristics, and the vulnerability and power of different actors.

• Institutions influence a person's social capital; the institutional arrangements that s/he is able to engage with affect the person's relative power within a community. A person's social capital and power determine her/his access to other assets and how much that person is able to gain from them.

• The development of institutional arrangements may reduce risk and vulnerability (eg, through the definition of property rights). Developing and maintaining institutional arrangements (eg, share-cropping contracts) are of critical importance to the poor.

4.2 METHODOLOGY

4.2.1 Study Sites

Thua Thien Hue is a coastal province, located in a key economic zone in Central Vietnam. It has a hydrology suited to the development of hydropower, with short rivers and steep slopes descending from the Truong Son Range, which runs through the province. The province has, therefore, been promoting the construction of hydropower facilities, including Binh Dien, Huong Dien, A Luoi, Ta Trach, A Roang, Hong Ha, Bo River, Thuong Nhat, A Lin, Hong Thuy, and Thuong Lo projects. Alongside these projects are five "displacement projects" worth a total investment of VnD 174,359 billion with 967 displaced HHs (Luyen, 2010).

Since 2005, with the construction of the Binh Dien Hydropower Plant, 51 HHs have been resettled to Bo Hon Village, comprising 241 people. Of these, 114 are classed as laborers; 45% of the population is considered impoverished. The village covers an area of 35 ha, and has 3 km of asphalted road, a primary school, a nursery station, a community hall, electricity supply for the whole village, and water supply derived from a nearby waterfall.

With the construction of the Ta Trach Dam in 2004, 797 HHs were resettled in seven communes. and 197 HHs have been resettled to Hoa Thanh (49 HHs), Binh Duong (74 HHs), and Hoa Binh (71 HHs) villages, comprising 858 people. The latter cover an area of 200 ha (165 ha cultivated and built-up areas; 35 ha of lake, rivulets, and uncultivated land). Infrastructure includes three community halls and 24 km of asphalted road.

4.2.2 Research Approach and Methods

4.2.2.1 Methods

The primary focus of this project's methodology is a "before and after" comparison to evaluate the impact of resettlement on PAPs. It focuses on changes to their livelihoods and NRA.

Conditions before the construction of the dams are derived from an already gathered data set, comprising data on area, types and quality of land, forest and water and property rights, and resource access rights; income, income-making activities, labor, work times, and the livelihood assets of people (before construction and at the time of resettlement).

To explore contemporary conditions, rapid rural appraisal (RRA) is used. RRA includes guidelines and tools, which help people to work in a structured but flexible way in the rural communities. RRA tools support efficient communication and interaction with those communities. Researchers can use these tools to collect data and answer research questions. RRA's participatory techniques help researchers to avoid setting research objectives based solely on their own perceptions, which often occurs with the traditional approach. The techniques that are used in this research are resource mapping, problem trees, seasonal calendars, Venn diagrams, focus group discussions, semistructured interviews, decision-making matrices, observations, and ranking.

The sustainable livelihood approach (SLA) (Scoones, 1998) is also used to evaluate the livelihoods, risks, vulnerabilities, and strategies of affected people within the context of changes to NRA. The SLA is "a way of thinking about the objectives, scope and priorities for development, in order to enhance progress in poverty elimination." It is based on the resources that people can access, and not their needs. It is also based on evolving poverty reduction theory, and thinking about how to break free of the decades of limited success with poverty alleviation policies (Ashley and Carney, 1999).

4.2.3 Data Collection

Primary data are derived from:

– Key informant interviews
– Household surveys (60 households)
– Mapping and resettlement area inventories
– Focus group discussions
– Stakeholder meetings

Secondary data was collected from Internet-based sources; commune, district, and provincial office reports; reports on the Binh Dien and Ta Trach compensation and resettlement programs; and reports of the Binh Dien Electric Company and other past research.

4.3 RESULTS AND DISCUSSION

4.3.1 Resettlement Program and Policies Associated With Dam Construction

Basically, the state requires that all of the compensative and resettlement programs have to ensure the benefits of people and the condition of living in the new place are the same or better than the old place, compensative level matches the value of damaged assets, land is compensated by land, and a house can be compensated by a house or money depending on negotiation between people and investors. The resettlement and compensation program of Binh Dien Hydropower and Ta Trach Dam were implemented correctly based on the rules of the state and province. However, difficulties with the implemented process always exist and lessons and experiences of the precious programs were shared. Difficulties of implementing policy and scheme were collected through a stakeholders meeting.

In the case of Ta Trach Dam, this situation also occurred. In the land compensation plan, a big area of forestland will be acquired from State Forestry Enterprises (SFEs) to compensate for affected HHs. The plan received agreement among SEFs, the Department of Natural Resources and Environment (DONRE), investors, the provincial people's committee (PPC) of Thua Thien Hue, and affected people. SEFs who are managing land forest in Binh Thanh Commune (host commune) will transfer land to DONRE, and after that DONRE will transfer land to the affected HH. However, this plan was broken in the implementation process because the process of land transfer between SEFs and DONRE had been delayed. Secondly, the spare land fund in Binh Thanh Commune is very small; therefore, the RA was allocated with a small area. Lastly, compensated land always attaches land-use change (LUC) to make sure the rights of an affected HH with land in a new place are protected as well as to avoid conflict in land use. However, the LUC issue process also relates to many stakeholders, and as a result process is delayed.

In the displacement process, the difficulty of both projects is the negotiation process with affected people to ensure the process will be implemented on time. However, even people who agreed and received full compensation, still did not want to move on to a new place per the plan, especially ethnic people in the Binh Dien Dam project. After many community meetings, the negotiation process defined the time to displace. Furthermore, when people already have moved on, some HHs still come back until their land and home in the old place are totally inundated.

Difficulties of compensation and displacement increase the pressure of livelihood support, which is the final step of the resettlement program. Furthermore, successful livelihood support strongly depends on the capacity of affected people while most of the people of both projects have a low education level and they only have cultivation skills on lowlands. People need to be supported in cultivation techniques on highlands or improve their skill to take a new nonagricultural job. To support people who can adapt and cultivate in a new place, the program needs to spend a lot of money over a long time frame. But the time for this step is only from 6 to 12 months.

Beside the difficulties that implementation has faced, the satisfaction of people with the compensation program and schemes also indicates the advantages and disadvantages of the policies. High satisfaction levels of the people mean the policies and schemes have met the needs of the people. When needs of the people problem are solved, implementation of policies is easier.

The satisfaction level of HHs about the resettlement program associated with dam construction is an index, which presents reasonable efficient aspects of policies and schemes. Based on the satisfied level of people, decision makers can revise policies and schemes to implement better and more efficient resettlement programs in the future. The satisfaction level is shown in Chart 4.1.

Basically, satisfaction levels of people in both RAs exist at two levels: satisfied and unsatisfied. These aspects of the resettlement program that received satisfied level more than an unsatisfied level are compensation for damaged assets, house compensation, explicit level of policies, technical training course, and development infrastructure with a rate of satisfied

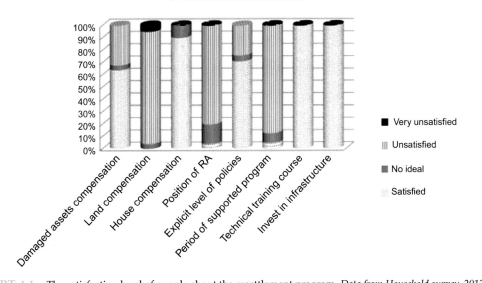

The satisfaction level of people about the resettlement program. *Data from Household survey, 2012.*

HH that varies from 60% to 100%. This reflects policies have been ensured equal on compensative price.

However, there are also aspects that do not respond yet to the desires of people such as land compensation, the position of the RA, and the food support program. The rate of unsatisfied HHs in these areas is high and varies from 80% to 91.2%, including 4.8% of HHs who are very unsatisfied with land compensation. Land compensation is a difficult issue that always exists in almost every resettlement program, especially programs associated with dam construction (Nha, 2012). The cause is the quality and area limitation of land resources in RAs. Land compensation and determination and the position of RAs are often allocated high land and small area. Unavoidably, people cannot accept it when they have to move from a low area with good land and big area to a high land with a high slope and bad quality. Some people have already come back to the old place. Furthermore, the transferred process land has been delayed and is unclear as the above discussion was made an urgent matter in the commune. People no longer have confidence in policies and plans that are drawn perfectly before they agree to be displaced. In addition, they only received food support for the first 6 months; after that people were let down with many difficulties and were shocked with serious change.

4.3.2 Impacts of Dam Construction on Natural Resource Access

4.3.2.1 Land Resource Access

Findings from the HH survey showed that there is a change in HH land access. In the compensation policy, land must be compensated with the same area of land, but Chart 4.2 shows that land area per HH after resettlement decreases sharply, from 2.92 ha/HH to 1.45 ha/HH in the BinhDien RA. Based on compensation documents, as well as data from the survey,

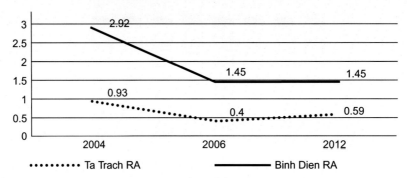

CHART 4.2   Change in land area per household (ha/HH). *Data from household survey, 2012.*

there are two reasons for this issue. In the BinhDien RA, there is not enough land to provide the same land area in compensation. Therefore, compensation for land was converted into a monetary value and disbursed. At the new location, people only have 0.5 ha/household. This area includes land for a house and garden. In the Ta Trach RA, all of the forestland that people were to receive in compensation still belongs to the SFE. After 7 years, the SFE still has not handed the forest-land over to the people.

Despite the decrease in land area, Chart 4.3 shows that land-use rights in the RAs are more clearly delineated than in the pre-resettlement site. Before resettlement, HHs were not issued certificates for the total area of land used, but after resettlement, certificates were issued for the total area of land used by each HH. The issuing of certificates for land use after resettlement is regulated under the compensation policy by the government.

However, the increase in land-use rights does not lead to increased efficiency of land use. This is showed in Chart 4.4, where the efficient use of both land types falls. In the Ta Trach RA, efficiency of cropland falls from 17.9 million Vnd/ha/year to 5 million Vnd/ha/year. Group discussion identified two causes for this problem. Almost all land was compensated with poor-quality rocky land on steep slopes. Another cause was the restriction of water access.

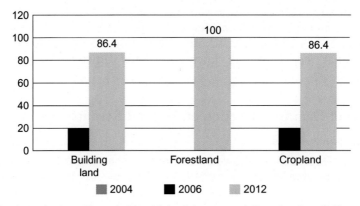

CHART 4.3   Change in the rate of households with land (percentage). *Data from household survey, 2012.*

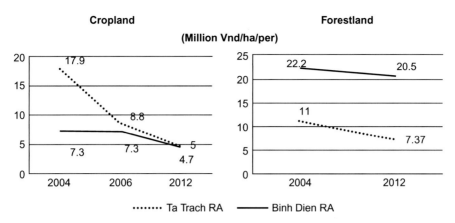

CHART 4.4   Change in land-use efficiency. *Data from household survey, 2012.*

After resettlement, water access of the HH reduces sharply. If the resettlement site is not close to a body of water, people use tap water, for which they must pay a fee. People then reduce their water usage to the minimum required for daily activities to avoid paying high fees. Agriculture uses groundwater from wells, but during the dry season, almost all of the wells are dry. Due to the lack of water, land quality improvement and cultivation declined.

4.3.2.2 Common Pool Resource Access

Common pool resources (CPRs) are resources that no individual has exclusive rights to such as fisheries, reefs, forests, pastures, and waterways. They are typically administered and owned by the social group, a village, or the state. Benefits from access to CPRs will be shared by the commune as a group rather than for individual people. Some decades ago, researches showed that HH income from CPRs was from 15% to 25%; for the poor, this rate was 29%. In rural India, income from CPRs annually is US$5 billion (WRI, 2005). In this research, CPRs are defined to include natural forest and water. Other CPRs that do not exist or exist with small amounts are ignored.

The quality and distance of access to water in both RAs has changed. Before resettlement, people lived beside the river; therefore, surface and groundwater was very near to their home. Moreover, these water sources were located upstream and had acceptable quality for both daily activities and cultivation. But after resettlement, the distance from home to a surface water source was greater and the depth to access groundwater increased.

Quality of water is also worse. Two reasons led to the increase of distance and decrease in quality. First, water pollution around the country has increased and these RAs are no exception. Second, since 2008 Binh Dien Dam has acted to block the water source that supports the people in RAs along Huu Trach River. This means both surface water and groundwater is no longer profuse. This issue becomes more stressful in the dry season from May to December.

In addition to water issues, Vietnam has about 25 million people who depend on forests (Thuan, 2005). According to Nguyen Sinh Cuc (2003), nontimber exploitation occupies 13.7% of total of HH income in rural areas. In areas where natural forest occupies a large part of the land, income from nontimber contributes a high rate to total income. However, compared with other income of HHs, income from nontimber exploitation is low. Beside its direct effect on HH income, access to forest also creates indirect impact. These impacts can improve the livelihoods of HHs who live nearby the forest (Thuan, 2005). This means that access to forest resources is very important in people's livelihoods. To evaluate forest access, we focused on rights, exploited frequency, and income from forest exploitation before and after resettlement.

In Binh Dien and Ta Trach areas both before and after resettlement, the types of natural forest that people can access are monotonous. These include community forest and protective forest. We focus on changing the right to access of people to natural forest. There are two types of forest that people can access before and after resettlement. Both after and before resettlement, right to access of people to protective forest is only the right to access and withdrawal. People only can exploit nontimber such as rattan, firewood, rush, and reed. However, before resettlement, production was very diverse and abundant. People could access and withdraw them easily. But from resettlement to the present, production of these exploited items has been poor or reduced because the forest that is near RAs was overexploited. There is not enough nontimber to exploit, so people have changed to the illegal exploitation of timber.

In terms of access to community forest, before resettlement, people in Binh Dien RA had one community forest with an area of 60 ha. This forest was planted in bamboo (Lo O). Upon becoming part of the RA, that community forest was acquired and compensated by money. Now, people cannot access it. With Ta Trach RA, before resettlement people could not access this forest, but now they have access to an area of 30 ha. They have planted acacia. Right to access of commune forest of people commonly is access and withdrawal, management and exclusion. However, the case in Ta Trach RA is not like that; people can access, management and exclusion but they do not allow withdrawal yet even though acacia can be exploited. The total area of commune forest in the Ta Trach RA belongs to a SFE. Before transfer of this area to the commune, the SFE did not harvest trees on the land. Therefore, the commune can plant trees, manage them, but are not allowed to exploit them. When the SFE exploits their trees, this area will belong to the commune, totally.

4.3.3 Impacts of Resettlement on Affected Households' Livelihoods

Rural livelihoods are dependent on natural resources, and as a result tend to be unstable and highly vulnerable to shocks in the supplies of natural resources. Livelihoods in RAs are more unstable, as people have to adjust to the change in living conditions. Diversifying income activities is one method used to adapt to displacement. Chart 4.5 shows the structure of income activities, before and after resettlement. After resettlement, the diversity of affected people's livelihoods increased to nine income activities. However, total average income per HH decreased from 27.25 million Vnd to 26.42 million Vnd, The key income activities are wage labor (38% of total income) and forest plantation (31% of total income).

Income is the final result of livelihood activities, but livelihood activities depend totally on livelihood assets. The quality and volume of assets determine the livelihood strategies of HHs, and also determine the level of importance each asset plays in contributing to HH income. The level of importance for each asset also shows livelihood of people is depending on which assets. In group discussion, a participant identified that three aspects of livelihood assets that impact on income are quality, scale, and profitability. After that, a participant compared and marked three aspects of each asset before and after resettlement. Ranking importance level is based on total marks from and reducing 1 to 5. The result of group

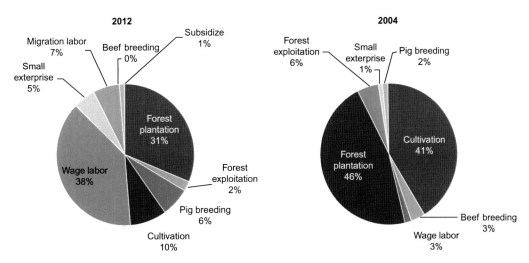

Change of income structure of households (percentage). *Data from household survey, 2012.*

discussion shown in Table 4.1 depicts change in the level of importance for assets. The key point is the change in the level of importance for natural, humane, and physical assets. When NRA decreases, its level of importance also declines. People switch their livelihood activities from depending on natural assets to depending on human assets (wage labor). However, this reduction is not by choice, and people do not have enough skills to engage in nonnatural asset-dependent livelihood activities. As a result, unemployment and social problems have been spreading. Data from HH surveys show that the number of workdays per laborer is only 142.5 days/year.

Vulnerability context is political, social, and economic environments where the livelihoods and lives of people are directly impacted. Vulnerability often suddenly occurs and fluctuates. Vulnerable assessment really needs to make a safe and sustainable livelihood strategy for the HH. In this research, we only include vulnerability that was occurred due to the resettlement and displacement program throughout group discussion. We used the tool of a problem tree to draw a picture about vulnerability. Look at the results in Fig. 4.1; we can see three risks that people have been facing: reduced food security, increased social evils, and increased female unemployment.

The first vulnerability is reduction of food security; there are many reasons leading to this issue. The main reason is reduction of NRA, particularly land resource. Generally, in rural areas and particularly in the RA, food security is dependent on land resource and other resources. Therefore, the loss of riceland and sownland means food security sharply reduces. The effort of quality improvement of gardenland to cultivate food crops such as bean, peanut, and cassava was a failure because of a lack of water and the high slope of the land. Although people were compensated by landgarden, its slope is high (>15°) and its quality is very bad. The first year when people moved here, they cultivated about 80% of the landgarden area, but only about 40% now. The area of uncultivated land has been increased; some HHs no longer cultivated land 5 years after resettlement.

TABLE 4.1 Change in Livelihood Assets' Contributions to Household Livelihood Income

Indexes	Before Resettlement					2011				
	N	P	H	S	F	N	P	H	S	F
Ta Trach resettlement area										
Quality	5	1	2	2	2	2	4	3	2	1
Volume	5	1	2	2	2	1	3	3	2	3
Profitability	4	1	4	1	3	3	2	5	1	3
Total marks	14	3	8	5	7	6	9	11	5	7
Ranking important level	1	5	2	4	3	4	2	1	5	3
Binh Dien resettlement area										
Quality	4	1	2	4	1	2	5	3	2	2
Volume	5	1	2	3	2	2	4	2	2	4
Profitability	5	1	3	3	2	3	2	5	1	3
Total marks	14	3	7	10	4	7	11	10	5	9
Ranking important level	1	5	4	2	3	5	1	2	5	3

Note: H, Human asset; N, Natural asset; P, Physical asset; F, Financial asset; S, social asset.
Marks reduce from 5 to 1; the best is 5, the worst is 1.
Data from group discussion, 2012.

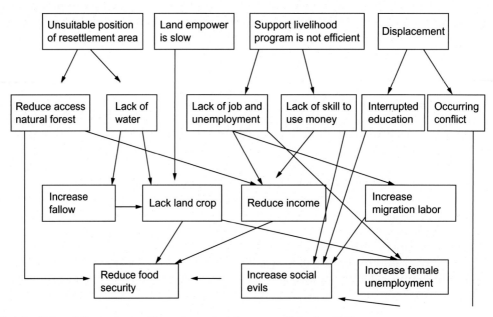

FIG. 4.1 Vulnerability in the resettlement area. *Data from group discussion, 2012.*

II. LAND USE SYSTEM AND LIVELIHOOD COMPLEXITIES

Besides, the decreased access to natural forest also leads to a reduction of food security, especially in Binh Dien RA where people often take some food such as bamboo shoot honey milk before resettlement. Additionally, food security could be ensured by income, but HH income also falls after resettlement. Lack of land crop and the lessening of CPR access and income while food support is so short, all put food security of affected HHs on an alert condition, especially with the poor HHs in both RAs.

"Before resettlement, we never had to worry about food which is always available, but now food is main interest of almost household, especially on the flood season when almost livelihood activities are interrupted. Almost men can take non-agriculture job but number of workday is a half of year, the salary from this source is enough to buy food in all year. Five year passed away but our life still is at a standstill and become more and more stressful" (Leader of farmer union in Ta Trach RA)

The second vulnerability is increase of social evils. Social evils are a complex issue and too different to solve. Social evils in both RAs include drug addiction, AIDS, thievery, school dropouts, and fighting. When people are compensated with money, some HHs use it to build a house or save it in the bank, but some HHs do not. Suddenly, they have a large amount of money and think they no longer need to work. Some have spent money to buy heroin and try it and as a result they become addicted. Besides that, displacement led to interrupted education of children. Adaptation to a new school is not easy for children, and their parents do not have enough time to care for them because they have to deal with difficulties at their new place. After resettlement a lot of children leave school. They have a lot free time and are not managed by their parents, so some of them become thieves to get money to play games or even buy wine to drink. Another issue is migration labor; some boys think they can get more money without working hard in the big cities. They moved to these cities, but they did not have enough skills to take new jobs as well as the skills to protect themselves from social evils. Up to now, most migration laborers have already come back to the RAs and unemployment. Many young people have already reached adulthood (>18 years) but still totally depend on their parents. Unlike the issue of food security, social evils increase in the well-off HH.

"Social evil is more and more complex and difficult to control. It is individual action of each people, we only can consult and advise them. Because their education level social knowledge are low, they fall in social evil easy when they move on the big cities. After that they come back RA and induce other young" (Leader of Young Union of Binh Thanh Commune)

The final vulnerability is the increase of female unemployment. This trend is more and more clear. Because of female health issues, they may not be a good match for some jobs; therefore, only some of them could participate in some employment activities. Some women have participated in small enterprises. Most of them are unemployed, and they are depending more and more on their husband or male labor in their family. This situation often leads to gender inequity in the commune.

> "we really feel uncomfortable with living in here despite house and infrastructures are better than old place because almost of us have been unemployed since moving on here. There are about 60% of women only stay at home and work housework. We really need to support to take new job at commune to gain initiative in our life" (Leader of women union of Ta Trach RA)

4.3.4 Impacts of Natural Resource Access and the Resettlement Program on Livelihood

The change of livelihood activities of people in RAs was caused by many things. These changes include both negative and positive factors. Impacts of the resettlement program on livelihood are expressed in Fig. 4.2. Basically, all components of HH livelihood were impacted directly or indirectly by the resettlement program.

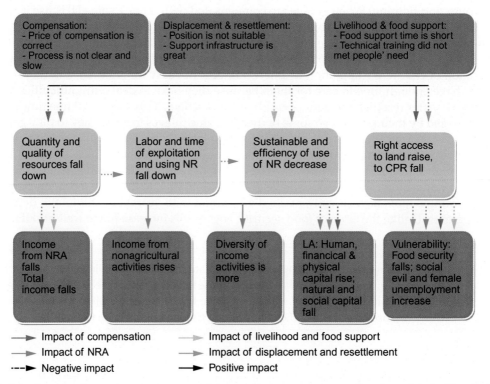

FIG. 4.2 Impacts of natural resource access and the resettlement program on affected people's livelihoods. *Data from group discussion, 2012.*

First, slow and unclear compensation, inefficient support of livelihood, and unsuitable position of RAs lead to reduced quantity and quality of NRA and result as income of HH from NRA reduce and vulnerability increase.

Secondly, income from nonagriculture increases but total income of HHs reduces; diversity of income activities are more than before; human, physical, and financial assets increase while social and natural assets reduce; vulnerability increases. The result of these change makes life of people more and more difficult.

4.4 CONCLUSION AND RECOMMENDATIONS

4.4.1 Conclusion

Hydropower development still has benefits for growing economies, which have a high demand for electricity. The increase in the number of hydropower projects leads to an increase in the number of HHs that are impacted both negatively and positively.

Displacement and resettlement lead to changes in the living conditions and livelihoods of affected HHs. The results of this study show that the NRA of affected HHs decreased. The underlying cause is the decrease in water resource access. Due to the lack of water, people cannot convert untouched land into cultivable land, nor can they engage in efficient cropping. The total land area per HH and efficiency of land use fell sharply. However, a positive impact is that the percentage of HHs with land-use certificates increased from zero before resettlement to 100% after. Despite this, land-use certificates do not seem to enhance people's quality of life, so most people do not perceive land-use certificates as increasing their land-use efficiency.

The decline in NRA leads to changes in the livelihood activities of affected HHs. The number of income activities is more diversified, but total income is lower than before resettlement. The income activity with the most participation is wage labor, which constitutes 38% of total income. This activity relies on human capital rather than natural capital. Hence the level of importance in terms of income generation from natural assets decreases, and the level of importance of human capacity increases. However, human capacity is limited by a lack of skills and education, which then leads to unemployment and increases in social problems.

4.4.2 Recommendations

Based on this study's findings, we recommend that displacement and resettlement associated with dam development should have its own compensation policy. It should link closely with water rights and access to make sure that there is equal access to water for affected HHs. Compensation and resettlement plans should be carefully prepared, especially when choosing where affected people will be resettled. The RA must have the same natural resource wealth as in the old location. If it does not, compensation should include not only access to land, but access to water as well. In addition, awareness of water access rights should be raised among all stakeholders, especially among affected HHs. This awareness will help ensure that compensation is calculated correctly and is adequate.

The compensation policy should require the hydropower owner to set up a fund to support affected HHs. This fund could be created through the designation of a small percentage of profits from the hydropower project. This fund will be used to support affected people in the long term, for 15–20 years. The fund and benefit-sharing scheme should be built with the participation of all stakeholders.

Additionally, studies on water use rights should be commissioned, particularly with respect to ways of measuring the value of water use rights. The results of these studies will be entered into a database to help decision makers allocate compensation equitably.

References

Ashley, C., Carney, D., 1999. Sustainable Livelihoods: Lessons From Early Experiences. Department for International Development, London, UK.

Chambers, R., Conway, G.R., 1991. Sustainable Rural Livelihoods: Practical Concepts for the 21st Century. Discussion Paper 296, Institute of Development Studies.

Cuc, N.S., 2003. Agriculture and Rural Areas of Vietnam in Reforming Period (1986–2002). Statistics Publisher, Hanoi.

Ellis, F., Allison, E., 2004. Overseas Development Group, University of Anglia, UK. Livelihood Diversification and Natural Resource Access. FAO, LSP WP 9, Access to Natural Resources Sub-Programme, Livelihood Diversification and Enterprise Development Sub-Programme.

Luyen, N.T., 2010. Problems of Implementation Process of Resettlement Policy for Ethnic Minorities in Dam Construction. Hue University, Thua Thien Hue province.

Ly, L.V., 2009. Displacement and Resettlement Policy for National Projects in Highland and Ethitc Areas—Problems Need to Solve. http//cema.gov.vn/modules.php?name=Content &op=details&mid=9264#ixzz19kI2eoZm.

Nga, D., 2010. Dam development in Vietnam: the evolution of dam-induced resettlement policy. Water Altern. 3 (2), 324–340.

Nha, P.D., 2012. Collection research results about lack of cultivation land and implementation land policies for minority ethnic in mountain area. CODE.

Pannture, 2008. Hydropower Trend in Vietnam: The Less Mentioned Social and Environmental Side-Effects. http//www.nature.org.vn/en/features/hydropower-trend-in-vietnam-the-less-mentioned-social-and-environmental-side-effects/.

Scoones, I., 1998. Sustainable Rural Livelihoods. A framework for Analysis. IDS Working Paper 72. University of Sussex, UK: Institute of Development Studies.

Tortajada, C., 2001. Environmental Sustainability of Water Projects. Envi Royal Institute of Technology, Stockholm, Sweden.

Thuan, D.D., 2005. Retrieved from http://www.wri.org/publication/content/8055.

Tuyet, D.T., 2011. Small dams a big cause for concern in Vietnam. Int. Water Power Dam Constr. 63, 18–22.

WB, 2011. Involuntary Resettlement. http//web.worldbank.org/WBSITE/EXTERNAL/TOPICS/EXTSOCIALDEVELOPMENT/EXTINVRES/0, menuPK410241~pagePK149018~piPK149093~theSitePK410235,00.html.

WCD, 2000. A new framwork for decision-marking.

Wri, W., 2005. Common Pool Resources as a Source of Environmental Income.

Yen, C.T.T., 2003. Towards Sustainability of Vietnam's Large Dams Resettlement in Hydropower Projects. Royal Institute of Technology, Stockholm, Sweden.

SCIENCE AND NATURAL RESOURCES MANAGEMENT

Structure and Diversity of a Lowland Tropical Forest in Thua Thien Hue Province

V.T. Yen*, R. Cochard†,‡

*Hue University of Agriculture and Forestry, Hue, Vietnam †Institute of Integrative Biology, Swiss Federal Institute of Technology, Zurich, Switzerland ‡Asian Institute of Technology, Klong Luang, Pathumthani, Thailand

5.1 INTRODUCTION

Central Vietnam is the most forested region of Vietnam (40.3% natural forest cover in Thua Thien Hue; MARD, 2015). The tropical forests are very rich in biodiversity and endemism. Overall 13,766 plant species have been recorded, and from new descriptions the list of species is still growing (MoNRE, 2011). The forests also harbor several species of large wildlife, including unique and threatened species, some of which (such as the mysterious saola, *Pseudoryx nghetinhensis*, discovered in Vu Quang Nature Reserve in 1992) have only been described in recent times (WWF, n.d.; Sterling et al., 2006). The forests provide essential ecosystem services to humans, including provisioning services (timber and nontimber products for food, fuel, and medicines) and important regulating services (eg, watershed protection, flood and landslide mitigation during storms, carbon sequestration) (Cochard, 2013). Despite this, the forests are increasingly threatened due to overexploitation and conversion to agricultural lands or plantation forests. Especially in the lowland regions, forests are cut at fast rates and are primarily replaced with plantations of fast-growing exotic tree species (predominantly *Acacia mangium* and *Acacia mangium×Acacia auriculiformis*) for the production of wood and pulp (Phuc et al., 2012; Sterling et al., 2006). The natural forests become more and more fragmented and the remaining forest patches are isolated, with plantation forests representing a barrier for the spread of most wild plant and animal species. Hence, threats of local and regional species extinctions are becoming ever more serious, particularly because other pressures on forests (eg, degradation via resource exploitation, exotic species invasions, and climate change) are intensifying (Dung and Webb, 2008; Cochard, 2011).

While the remaining forests in the lowlands are disappearing and being degraded at a fast rate, detailed scientific documentation of forests' original "natural" species wealth and vegetation structure are still few. Different types of lowland vegetation have not been described in much detail, and data on descriptive parameters of lowland forests such as tree density, basal areas (BAs), floristic richness, and diversity are often lacking (Dung, 2010; Blanc et al., 2000). Here we report on the structure, biomass, and species composition of an isolated lowland evergreen rain forest fragment (surrounded by acacia plantations). Like in most forests in the Central Vietnamese lowlands large timber trees (*Hopea pierrei*, *Erythrophleum fordii*, and *Barringtonia macrostachya*) had been strongly logged during the 1990s (Detective-90, 1992), so that the studied forest may no longer be described as pristine. Nonetheless, the forest structure and species composition was maintained to a large degree, and the forest vegetation may be fairly representative of natural forests in the region, where many persist in a somewhat degraded state. Documentation of the forest structure and floristics will be key to the further study of forest dynamics, forest management, and effective biodiversity conservation (especially in isolated forest fragments). The objectives of this study were to (1) survey the plant species composition and forest structure and (2) assess the plant biodiversity value of the forest in terms of species richness and using various biodiversity indices.

5.2 STUDY SITE

Nam Dong is a remote, mostly hilly to mountainous district in Thua Thien Hue Province in Central Vietnam. It is characterized by high mean annual precipitation of about 2500 mm, and its forests are known to harbor high plant diversity (Tordoff et al., 2003; Dung and Webb, 2008). Forest cover in Nam Dong District has declined and become fragmented due to logging, infrastructure development, and overexploitation of timber and nontimber forest products.

The study site was a 42 ha large forest fragment located in the southeast of Nam Dong District belonging to Huong Phu Commune in the Huong River watershed area (Fig. 5.1; 16°00'–16°15'N; 107°27'–107°53'E). The forest was a lowland evergreen rain forest, which was, however, located on a sloping hill at an elevation of between 40 and 320 m above sea level, with drier parts of the forest on steep hillsides (inclinations up to >45 degree) and on hill ridges, and more shady, moister parts at valley bottoms near permanent creeks (Fig. 5.1). The forest is mostly regenerating natural forest since logging has occurred in this compartment. Major timber trees are *B. macrostachya*, *Erythrophleum fordii*, *Knema conferta*, *Gironniera subaequalis*, *Castanopsis* spp., *Ormosia* spp., and *Syzygium* spp. (Dung, 2010).

5.3 METHODS

5.3.1 Data Collection

Forty 20×20 m square plots (400 m²) were established in the forest: 24 plots were located on hill ridges (covering 0.96 ha in total) and 16 plots were located close to streams (0.64 ha). All plots were at least 30 m apart. Tree bole diameter at breast height (DBH) was measured

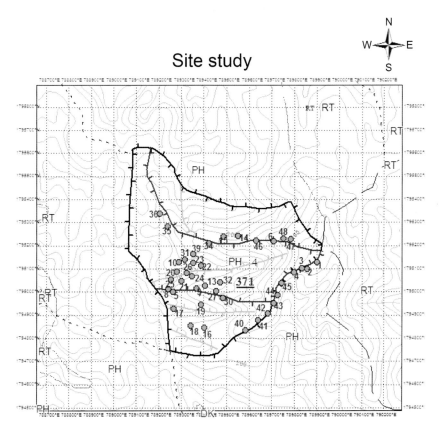

FIG. 5.1   Study site at Huong Phu Commune. The plot locations are indicated by green dots.

1.4 m above ground. All dicotyledonous mature trees (DBH >6 cm; as defined by IFRI, 2004) and tree saplings (DBH 2.5–6 cm) within the plots were counted, whereby the species were identified by comparison to a local herbarium collection, and using available plant identification literature (Ho, 1999; FIPI, 1996). For proper species identification images were made (of leaves, bark, etc.) in the field with an electronic camera (information was gathered in a database), and plant material was collected and dried in a plant press for comparison and identification at the herbarium. The bole diameters of all recorded trees were measured with a girth tape, and the height of the trees was identified using an inclinometer and a 50 m tape (one person shaking the tree, so that the top leaves could be seen moving through the forest canopy when measuring). The densities of seedlings (DHB <2.5 cm or height <1.4 m) within plots were determined from five 2×2 m squared subsamples (20 m² in total), whereby four subsamples were located in the corners (clockwise A, B, C, D) and one (E) in the midpoint of the 400 m² main plot.

In addition, assess to light penetration onto plots and seedling recruitment canopy cover (%) and ground cover (%) was described. To identify canopy cover, we used a camera to take photographs in each subplot and determined the percentage of sky cover by using Photoshop. Herb cover was assessed visually.

5.3.2 Data Analyses

5.3.2.1 Basic Data Calculations

From the collected data, tree population densities and mean canopy heights were determined for all plots. In addition, further variables that describe vegetation structure and diversity were calculated; that is, tree BA, relative density, relative frequency and relative dominance, and importance value index (IVI) (Cottam and Curtis, 1956; Phillips, 1959):

- BA in m^2: BA = JI*(bole diameter)^2/4
- relative density = (total tree count of a species)/(total tree counts of all species) × 100
- relative frequency = (frequency of species on plots)/(total number of species) × 100
- relative dominance = (total BA of a species)/(total BA of all species) × 100
- IVI = relative density + relative frequency + relative dominance.

The total aboveground biomass (TAGB) of dry-matter woody trees was assessed using a formula by Brown et al. (1989): $TAGB = e^{[-3.1141 + 0.9719 * \ln (DBH^2 \times H)]}$.

Excel 5.0 and Biodiversity Professional 2.0 were used to estimate biodiversity levels of all species in this study, particularly using indices (Magurran, 2007; Odum, 1971):

- Shannon-Weiner index $(H') = -\Sigma p_i \ln p_i$.
- Simpson's index (D): $D = \sum_{i=1}^{n} \dfrac{n_i (n_i - 1)}{N(N-1)}$

In the formula, n_i represents the number of individuals in the ith species, N the total number of individuals, and p_i equals n_i/N. As D increases, diversity decreases. Therefore, Simpson's index is usually expressed as $1-D$ or $1/D$. Simpson's index is influenced by the most abundant species in the sample and is less sensitive to species richness (Magurran, 2007).

5.4 RESULTS

5.4.1 Forest Structure

5.4.1.1 Tree Density and Basal Area

Mean density of mature trees on plots was $619 \pm 167 \, ha^{-1}$ (range 328–1031 ha^{-1}), whereas density of tree saplings was $278 \pm 211 \, ha^{-1}$ (range 25–1150 ha^{-1}). Mean BA for all tree stems was $30.0 \pm 9.3 \, m^2 ha^{-1}$ (range 12.4–52.5 m^{2}ha^{-1}).

Data of tree DBHs were log-normally distributed. Average overall tree diameter was 15.1 ± 6.8 cm (for mature trees 18.2 ± 6.4 cm) with a range of 6–89 cm (Fig. 5.2). Only 0.12% of trees had a DBH >60 cm. The largest tree (DBH of 89 cm) was an individual of *Canarium bengalensis*. More than half of the trees (62%) were between 5 and 10 m tall, whereas a few (1.3%) were larger than 20 m (mean height 9.1 ± 2.7 m; range 1.6–25.0 m). The species of the tallest individuals were *Endospermum chinense* (25 m) and *Artocapus rigidus* (24 m).

5.4.1.2 Tree Biomass

The estimated mean TAGB of all trees on the plots was $144 \pm 64 \, Mg \, ha^{-1}$ (range 32–273 Mg ha^{-1}). Overall biomass was higher on the ridges ($159 \pm 73 \, Mg \, ha^{-1}$) than near the

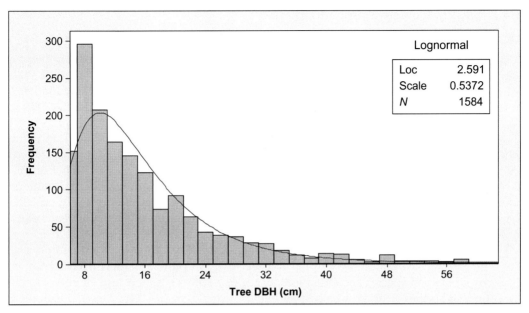

FIG. 5.2   Frequency distribution (and lognormal fit) of all measured tree DBHs at the Huong Phu forest.

stream ($121 \pm 41 \, \text{Mg ha}^{-1}$), with greater differences among saplings than for mature trees (Fig. 5.3). The potential TAGB assessed for tree stumps (ie, trees cut during logging) amounted to $11.2 \pm 0.8 \, \text{Mg ha}^{-1}$, and including this figure TAGB increased to around $155 \pm 63 \, \text{Mg ha}^{-1}$. Therefore, in terms of biomass removal, there was approximately 7% of biomass lost.

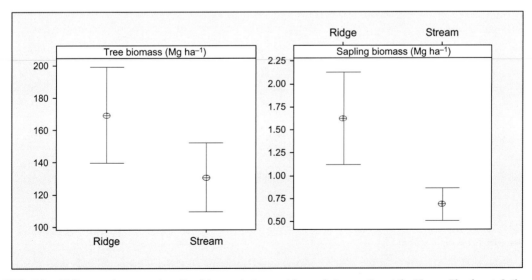

FIG. 5.3   Differences in tree and sapling biomass between ridge and stream sites at the Huong Phu forest. Spikes are the standard error. The *t*-test *p*-values indicate the statistical significance of the differences.

5.4.2 Species Composition and Diversity

5.4.2.1 Species Richness

A total of 154 tree species (12 unidentified) were recorded belonging to 45 families, with 134 species of mature tree species and 88 species of saplings. Families with the most species were Lauraceae (18), Euphorbiaceae (15), Moraceae (9), Sapindaceae (7), and Sterculiaceae (7), Burseraceae (4), Clusiaceae (4), Elaeocarpaceae (4), Fagaceae (4), Anacardiaceae (3), Fabaceae (3), Myrtaceae (3), and Sapotaceae (3) (Appendix A). Species composition in plots varied, and none of the species were found in all plots. However, there were 13 species found within more than 20 plots, and 2 species were represented by a large number of trees: *B. macrostachya* (52 cm in DBH, Lecithidaceae) and *Lithocarpus amygdalifolius* (63 cm in DBH, Fagaceae). Table 5.1 illustrates the difference in species richness, number of individuals, and number of families between ridge and stream plots. Tree species richness was 145 for ridge and 135 for the stream plots. Ridge plots were more diverse at all taxonomic levels than stream plots in terms of mature tree species richness and sapling species richness (p-value < 0.05, t-test).

The species-area curve in Fig. 5.4 illustrates the continually increasing number of tree species with the increasing number of plots surveyed (ie, 40 plots totaling 1.6 ha). There were differences between species accumulation curves from plots situated near rivers and ridges showing that species richness was overall higher near ridges ($p < 0.05$, t-test), but flattening out somewhat in comparison to the curve of the stream sites.

TABLE 5.1   Summary of Variables Describing Forest Structure and Diversity

Description of Variables	Ridge (24 plots)	Stream (16 plots)	Total (40 plots)
Forest structure	Average (and range in brackets)		
Mature tree density (ha^{-1})	1090 (550–1650)	841 (525–1175)	990 (550–1650)
Tree sapling density (ha^{-1})	349 (25–1150)	172 (25–350)	278 (25–1150)
Tree seedling density (m^{-2})	1.42 (0.4–2.1)	1.01 (0.4–2.6)	1.26 (0.4–2.6)
Tree basal area (m^7 ha^{-1})	32.1 (12.4–52.4)	26.8 (16.2 35.0)	27.0 (12.4–52.4)
Est. tree woody biomass (t ha^{-1})	159 (40.1–272.7)	121 (32.1–194.5)	144 (32.1–272.7)
Tree height (mature and saplings) (m)	8.3 (2–25)	8.2 (2–23)	8.3 (2–15)
Maximum tree height (m)	18 (21–25)	18 (11–23)	18 (11–25)
Tree crown cover (%)	81 (66–92)	81 (73–93)	81 (66–93)
Herb ground cover (%)	49 (27–64)	55 (36–79)	50 (27–79)
Forest diversity	Total on all plots		
No. of tree species (mature and saplings)	145	137	157
No. of tree families	37	32	45
Shannon-Wiener index (H')	2.77	3.00	2.91
Simpson index (1−D)	0.95	0.95	0.95
No. of seedling species	49	37	53

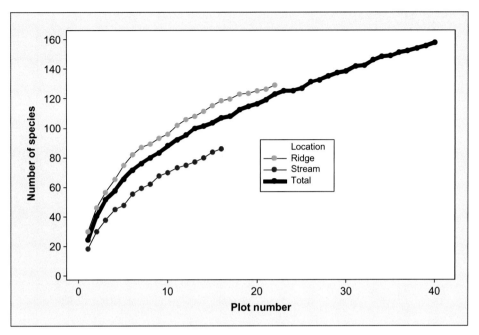

FIG. 5.4   Tree species-area curves in plots at the stream and ridge locations and in total at the Huong Phu forest.

5.4.2.2 Patterns of Tree Species Abundance

Fig. 5.5 describes the relationship between the number of tree and sapling species and the relative number of individuals in the 40 samples. It shows that the vast majority of the species of trees and saplings were relatively rare; more than 80 species were represented by only 1–3 species. Approximately 38.9% of tree species and 45.1% of sapling species were represented by only one individual. A few species were more common; for the trees, the most abundant species represented about 5.5% of all individuals, whereas for the saplings the most abundant species represented about 10%.

5.4.2.3 Species Composition in Regard to Forest Structure

Roughly two canopy layers could be identified with a somewhat differing mixture of tree species. The upperstory layers between 15 and 25 m in height mostly consisted of the tree species *Scaphium lychnophorum*, *Endospermum chinense*, *Artocarpus rigidus*, and other tall trees (Fig. 5.6). In contrast, the intermediate stratum was mainly composed of *Gironniera subaequalis* and other species such as *Lithocarpus amygdalifolius*, *Castanopsis* spp., and *Palaquium annamense*. The understory growth (5–15 m) was dense. It was mostly composed of *Gironniera subaequalis*, *Artocarpus rigidus*, and other species.

5.4.3 Natural Regeneration

A total of 987 tree seedlings belonging to 53 species were recorded on the 200 subplots (800 m²) during the entire survey; by extrapolation, tree seedling densities therefore ranged

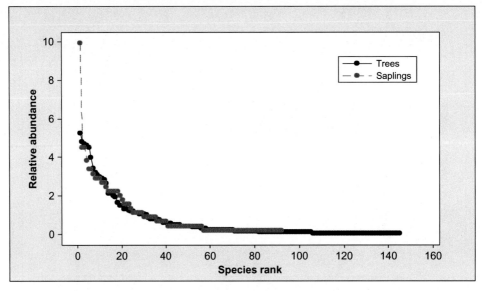

FIG. 5.5　Tree and sapling species rank/abundance diagrams for 40 samples at the Huong Phu forest.

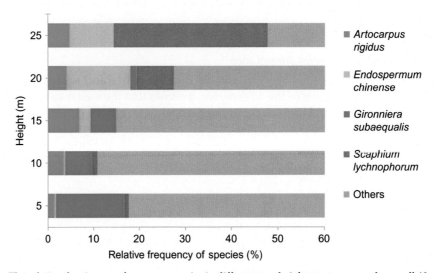

FIG. 5.6　The relative dominance of common species in different tree height strata summed over all 40 study plots at the Huong Phu forest.

from about 4000 to 27,000 individuals ha⁻¹ (average $13{,}000 \pm 6000$ ha⁻¹). *Artocarpus rigidus* was the species with the highest seedling abundance (79 trees and saplings) and regeneration with 148 seedlings. In contrast, the number of mature tree individuals of *Gironniera subaequalis* was high (127 trees and saplings), whereas few seedlings (42 seedlings) were counted. Seedling counts on ridges were higher as compared to stream plots.

5.5 DISCUSSION

5.5.1 Stand Structure

The forest had a high tree density compared to other tropical rainforests (Table 5.2). The density of mature trees ($\geq$10 cm DBH) in our study (678 trees ha^{-1}) was lower only compared to a site in Sarawak (739 ha^{-1}), but it was comparable to values recorded at sites in Sabah and India. In contrast, tree BA in this study was comparatively low with only 29.5 m^{2}ha^{-1}. This value was comparable to recorded BA values at sites in Vietnam (Cat Tien NP), China, and in the Bolivian Amazon. Again, the highest values (57 m^{2}ha^{-1}) were recorded for the site at Sarawak (Table 5.2).

Forests with similar canopy structures have been described in tropical forests especially in Brunei and Sarawak (Ashton and Hall, 1992) and in Uppangala, India (Pascal and Pelissier, 1996). The distinctions of structural forests were mainly attributed to differences in the available water in the soil or the nutrient content of the soil, which affect the main canopy (Pascal and Pelissier, 1996).

5.5.2 Tree Biomass

The biomass of Huong Phu secondary forest (144 Mg ha^{-1}) is comparable to other secondary or disturbed forests (Table 5.3). In nature, the biomass of primary forests is typically higher than in secondary forests due to biomass removal by disturbances.

TABLE 5.2   Comparison of Mean Density, Basal Area, and Species Richness of Trees $\geq$10 cm DBH in the Present Study at Huong Phu Forest, Central Vietnam, Compared With Some other Tropical Forests

Location	Plot Area (ha)	Minimum GBH (cm)	Mean Density (ha^{-1})	Mean Basal Area (m^{2}ha^{-1})	Species Richness	References
Sepikok, Sabah	1.81	30.5	660	42.1	198	(1)
Mulu, Sarawak	1	31.4	739	57	214	(2)
Sungei-Menyala, Malaysia	2	31.4	493	32.4	244	(3)
Uppangala, India	3.12	30	635	39.7	91	(4)
Cat Tien Vietnam, Plot E[a]	1	31.4	469	31.3	81	(5)
Nam Dong, Central Vietnam[a]	2.89	30	726	40.6	126	(6)
Xishuangbanna, China	0.8	31.4	386	30	77	(7)
Bolivian Amazon[a]	1	30	544	29.7	81	(8)
Huong Phu[a]	1.6	30	678	29.5	127	(9)

The minimum girth at breast height (GBH) is an artificial measure that slightly differed among studies. References: (1) Nicholson (1965), (2) Proctor et al. (1983), (3) Manokaran and Kochumen (1987), (4) Pascal and Pelissier (1996), (5) Blanc et al. (2000), (6) Dung (2010), (7) Cao and Zhang (1996), (8) Marielos (2001), (9) Present study.
[a]*Secondary forest.*

TABLE 5.3   Estimation of Biomass in Various Tropical Regions

Forest Types	TAGB (Mg ha^{-1})
Cameroon Tropical Moist Forest (primary forest)	279.2
Sri Lanka Tropical Moist Forest	
Primary forest	177.3
Secondary forest	124.9
Malaysia Tropical Moist Forest	
Disturbed Hill forest 1	156.3
Disturbed Hill forest 2	210.0
Huong Phu secondary forest (present study)	144.0

Source: Brown et al., 1989.

From the still visible traces of tree stumps or dead trees, it was estimated that approximately 7% of total biomass was lost within the plots in recent times. The main reason was logging activities as evident from many tree stumps found in the forest, especially along the stream area. We recognized very few dead trees due to diseases (insects), storms, and so forth.

5.5.3 Diversity

5.5.3.1 *Tree and Sapling Composition*

Huong Phu is a natural forest with disturbances mainly due to activities like logging and changes occurring from isolation due to land encroachment by surrounding acacia plantation forestry. This study may play a major role in conservation of plant species and management of the forest. The most common tree families were Lauraceae, Euphorbiaceae, and Moraceae, which is comparable to previous research in Vietnam such as Dung (2010), Fordjour et al. (2009), Hadi et al. (2009), Averyanov et al. (2006), Rundel (1999), Loc (1998), and Thin (1994). Although Dipterocarpaceae were considered to be a numerically dominant family in Southeast Asia by Whitten (1982), accounting for up to 22% of species, only three species were found in our study and five species in Uppangala (India) by Pascal and Pelissier (1996). The present study rather adds support to the suggestion that Lauraceae, Euphorbiaceae, and Moraceae are dominant families in tropical rainforests of southern Asia (cf. Padalia et al. 2004). The dominant species found in Huong Phu, in particular *Artocarpus rigidus*, *Gironiera subaequalis*, and *Castanopsis* spp., were also species reported to be dominant in Vietnam by Rundel (1999), PKKNP (2009), and Dung (2010).

On the whole, floristic richness of the Huong Phu secondary forest (157 species in 1.6 ha) was higher than in many other moist semideciduous forests. For instance, Anning et al. (2008) and Fordjour et al. (2009) reported much lower species richness with 59 species on 1.6 ha and 77 species on 1.6 ha in Ghana, India. These were, however, deciduous forest types where lower numbers of species are generally found than in tropical forests, even if forests are disturbed. Moreover, compared to our study the number of species was quite low in Phu Khao Khouay National Park in Laos, with only 117 species in 2.5 ha (PKKNP, 2009). In this study, even at the higher diameter (≥ 10 cm), the number of species (127 species in 1.6 ha) recorded was still higher compared to 91 species recorded in 3.12 ha of tropical evergreen forest in Uppangala, India (Pascal and Pelissier, 1996).

This study was conducted close to sites in Nam Dong forest (Dung, 2010) with 101 species in 1 ha. Additionally, the floristic composition (≥10 cm DBH) in Huong Phu forest is similar to other research such as 198 species in 1.81 ha in Sabah (Nicholson, 1965), 77 species in 0.8 ha in China (Cao and Zhang, 1996), and approximately 81 species in 1 ha in Cat Tien Vietnam (Blanc et al., 2000) and in the Bolivian Amazon (Marielos, 2001). On the basis of only trees (DBH ≥10 cm), the diversity is lower as compared to studies from the wet tropics of Sarawak with up to 214 species in only 1 ha (Proctor et al., 1983) or 244 species in 2 ha in Malaysia (Manokaran and Kochumen, 1987).

The number of species recorded in this study was comparatively lower than other similar studies in other tropical rain forests. For example, Campbell et al. (1986) reported 189 species ha^{-1} in Brazilian Amazon and Riswan (1987) recorded 160 species ha^{-1} in Indonesia. As we know, the Amazon forest is one of the forests with the highest numbers of species and it is a primary forest as well. Furthermore, the secondary forests were established from primary forest because of logging, including impacts such as collection of fuel wood and nontimber forest products and invasive species. In contrast, the secondary forest after several decades can become primary forest without any disturbances, provided the overall species pool is still intact in the forest. Generally, however, secondary forests have a lower number of species than the primary forest.

Comparing the ridge and stream sites, species richness in ridge plots was higher than in stream plots with 145 species and 135 species, respectively. Evaluating species accumulation curves, Hadi et al. (2009) found in evergreen lowland forests in Indonesia that species richness was slightly lower along river streams as compared to ridge areas. However, this result differed from a study of Mueller-Dombois and Ellenberg (1974 in Barbour et al. 1987) in tropical rain forest in Brunei. They reported that the species richness in valley bottom (stream) areas was much higher than in ridge areas with 150 and 60 species, respectively. Explanations may be that high species richness closer to the ridge in Huong Phu forest was due to higher disturbances in the stream areas such as logging by local people.

5.5.4 Natural Regeneration

5.5.4.1 Tree and Seedling Diversity

The density of seedlings (1.3 m^{-2}) was low in comparison to other studies. For example, Dung (2010) reported seedling densities of around 4 m^{-2}. Plot location and the elevation were probably important for seedling survival, and tree density and crown cover may also have affected seedling establishment and early survival. In stream areas high soil humidity and good conditions for growth of grasses and herbs may have diminished seedling establishment.

There were many species that were present in the tree layer but absent in sapling and seedling layers such as *Enicosanthellum plagioneurum*, *Garcinia* spp., *Croton* spp., *Cinnamomum* spp., and *Actinoda phnepilosa*. However, these species were recorded in low numbers as seedlings, and hence in the future these species may disappear from the upper layers. However, some dominant tree species were still represented in the sapling and seedling layers such as *Artocarpus rigidus*, *Gironnneira subaequalis*, *Garcinia cochichinensis*, *Knema pierei*, and *Syzygium syzygioides*. In contrast, even though *Gagura pinnata, Castanopsis* spp., and *Litsea sebifera* were very common in the tree layer, these were still very low in sapling and seedling layers. Therefore, at later stages, these species might no longer be dominant and may be replaced by other common species in sapling and seedling layers such as *Antidesma hainanense, Croton tiglium, Dillenia scabrella, Euodia lepta, Glochidion eriocarpum, Melanorrhoea laccifera, Pometia* spp., *Pometia pinnata*, and others.

5.5.4.2 Tree and Sapling Diversity

The diversity of mature trees was higher compared to sapling diversity. This is partly due to the greater number of mature tree individuals and species, which again is influenced by how the mature/sapling tree distinction was made. The diversity of saplings near the ridge was higher than near the stream. This can be explained by the higher density of saplings near the ridge compared to the stream. Another reason, logging activities in the stream may also be a reason for decreased plant diversity in this area. In ecological aspects, stream areas have often-higher climber, grass, and herb cover than ridge areas. The higher diversity of saplings (H') near the ridge can be attributed to higher sunlight intensity and more open gaps and other good conditions for saplings development (Fordjour et al., 2009).

High-value timber species were absent or found only at very low density in Huong Phu forest; this was due to logging of these species, which were in high demand. For instance, *H. pierrei* is one of the highest-valued timber species with high quality of wood. This species was present as only one individual as each mature tree, sapling, and seedling in the 1.6 ha study area. Other species, *Peltophorum tonkinensis, Madhuca pasquieri, Sindora tonkinensis*, and *Sindora siamensis* were present with a few individuals and this was also the case in the research by Dung (2010). In addition, Dung (2010) found that *Erythrophleum fordii* and *Tarrietia javanica* with high quality of timber in Nam Dong forest were absent even for seedlings, which was the same in our study. The removal of mature tree species with high timber value does not only affect sapling species, but also influences the seedlings (Toledo-Aceves et al., 2009; Lobo et al., 2007). As a result, the diversity of trees may remain high but it does not mean that a high timber value is being maintained in the forest. According to Slik et al. (2004), selective logging might be easy for early succession species, which are fast growing in the gap area. At a certain time period, the growth of old forests can have little diversity and is characterized by higher abundances of range-restricted species.

5.6 CONCLUSIONS

5.6.1 Species Composition and Forest Structure

Lauraceae, Euphorbiaceae, and Moraceae were the most common families in this forest. *Endospermum chinense* and *Gironniera subaequalis* were found to be dominant species on ridge plots and *Scaphium lychnophorum, Garuga pinnata*, and *Antidesma hainanense* were dominant on stream plots. *Artocarpus rigidus* was one of the most common species at the seedling level. The numbers of seedlings on ridge plots were higher than on stream plots. The density of trees, saplings, and seedlings increased with altitude. The ridge plots had higher tree BAs than the stream plots. In terms of height, the forest structure was divided into two kinds of structure in this forest, which included the upperstory layer (15–25 m) and understory layer

(<15 m). Low structures were found along the streams with low slopes and tall structures in gentle slopes or on the ridges. Huong Phu secondary forest is considered a young forest that was logged and on the rims encroached by commercial trees planted by the local people (particularly in the lower area—stream areas). Huong Phu forest not only has a rich floristic composition but also a high tree density. However, the forest resources in this area have been threatened by disturbance factors, especially along the streams.

5.6.2 The Biodiversity Value of the Forest

Species richness increased with altitude. The number of species near ridge plots was higher than in plots along the stream. The pattern of tree species abundance of secondary forest vegetation in Huong Phu is very unequal, presenting a long tail at the end of the curve in the rank/abundance diagram. The tree species diversity indices of plots near the ridge were not significantly different from stream of forest. However, they were higher in sapling species diversity indices.

There are numerous species with small population sizes, particularly in the species represented by only one individual, which can influence the tree species diversity value. *H. pierrei*, *Peltophorum tonkinensis*, *Madhuca pasquieri*, *Sindora siamensis*, and *Sindora tonkinensis* are trees found in this forest, all of which have very high wood quality.

References

Anning, A.K., Akyeampong, S., Addo-Fordjour, P., Anti, K.K., Kwarteng, A., Tettey, Y.F., 2008. Floristic composition and vegetation structure of the KNUST Botanic Garden, Kumasi, Ghana. JUST. Int. J. Biodvers. Conserv. 28, 103–116.

Ashton, P.S., Hall, P., 1992. Comparisons of structure among mixed dipterocarp forests of North-Western borneo. J. Ecol. 80 (3), 459–481.

Averyanov, L.V., Loc, P.K., Hiep, N.T., Harder, D.K., 2006. An Assessment of the Flora of the Green Corridor Forest Landscape, Thua Thien Hue Province, Vietnam. Report no. 1: Part One. Hanoi. Green Corridor Project, WWF Greater Mekong and Vietnam Country Programme and FPD Thua Thien Hue Province.

Barbour, M.G., Burk, J.H., Pitts, W.D., 1987. Terrestrial Plant Ecology. The Benjamin/Cummings Publishing Company, Menlo Park, California.

Blanc, L.G., Maury, L., Pascal, J.P., 2000. Structure, floristic composition and natural regeneration in the forests of Cat Tien National Park, Vietnam: an analysis of the succession trends. J. Biogeogr. 27, 141–157.

Brown, S., Gillespie, A.J.R., Lugo, A.E., 1989. Biomass estimation methods for tropical forest with applications to forest inventory data. J. For. Sci. 35 (4), 881–902.

Campbell, D.G., Daly, D.C., Prance, G.T., Maciel, U.N., 1986. Quantitative ecological inventory of terra firme and Varzea tropical forest on the Rio Xingu, Brazilian Amazon. Brittonia 38 (4), 369–393.

Cao, M., Zhang, J., 1996. Tree species diversity of tropical forest vegetation in Xishuangbanna, SW China. J. Biodivers. Conserv. 6, 995–1006.

Cochard, R., 2011. Consequences of deforestation and climate change on biodiversity. In: Trisurat, Y., Shrestha, R., Alkemade, R. (Eds.), Land Use, Climate Change and Biodiversity Modeling: Perspectives and Applications. IGI Global, Hershey, USA, pp. 30–55. Chapter 2.

Cochard, R., 2013. Natural hazards mitigation services of carbon-rich ecosystems. In: Lal, R., Lorenz, K., Hüttl, R.F., Schneider, B.U., von Braun, J. (Eds.), Ecosystem Services and Carbon Sequestration in the Biosphere. Springer, Heidelberg, pp. 221–293. Chapter 11.

Cottam, G., Curtis, J.T., 1956. The use of distance measurement in phytosociological sampling. J. Ecol. 37, 451–460.

Detective-90, 1992. Retrieved from: http://thuvienphapluat.vn/van-ban/Bat-dong-san/Nghi-dinh-90-CP-Quy-dinh-den-bu-thiet-hai-khi-Nha-nuoc-thu-hoi-dat-de-su-dung-vao-muc-dich-quoc-phong-an-ninh-loi-ich-quoc-gia-loi-ich-cong-cong-38851.aspx.

Dung, N.T., 2010. Exploring conditions for sustainability of community forest management in Nam Dong district, Central Vietnam. Thesis submitted in partial fulfilled of the requirement for the degree of Doctoral of Science, Asian Institute of Technology. School of Environment, Resource and Development, Thailand, December, 2010.

Dung, N.T., Webb, E.L., 2008. Combining local ecological knowledge and quantitative forest surveys to select indicator species for forest condition monitoring in central Viet Nam. Ecol. Indic. 8, 767–770.

FIPI (Forest Inventory and Planning Institute), 1996. Vietnam Forest Trees. Agricultural Publishing House, Vietnam.

Fordjour, P.A., Obeng, S.A., Anning, A.K., Addo, M.G., 2009. Floristic composition, structure and natural regeneration in a moist semi-deciduous forest following anthropogenic disturbances and plant invasion. Int. J. Biodivers. Conserv. 1 (2), 21–37.

Hadi, S., Ziegler, T., Waltert, M., Hodges, K., 2009. Tree diversity and forest structure in northern Siberut, Mentawai Islands, Indonesia. J. Trop. Ecol. 50, 315–327.

Ho, P.H., 1999. An Illustrated Flora in Vietnam. Parts I, II, III, Young Publishing House, Ho Chi Minh City (Vietnamese).

IFRI (International Forestry Resources and Institutions), 2004. Asian Institute of Technology. Thailand.

Lobo, J., Barrantes, G., Castillo, M., Quesada, R., Maldonado, T., Fuchs, E.J., Solís, S., Quesada, M., 2007. Effects of selective logging on the abundance, regeneration and short-term survival of Caryocarcostaricense(Caryocaceae) and Peltogynepurpurea(Caesalpinaceae), two endemic timber species of southern central America. For. Ecol. Manage. 245, 8–95.

Loc, P.K., 1998. On the systematic structure of the Vietnamese flora. In: Floristic Characteristics and Diversity of East Asian Plants: Proceedings of the First International Symposium on Flora Characters and Diversity of East Asian Plants. Kunming. 1996. China Higher Education Press, Springer Verlag, Beijing, Berlin, pp. 120–129.

Magurran, E.A., 2007. Measuring Biological Diversity. Blackwell Publishing, Oxford.

Manokaran, N., Kochumen, K.M., 1987. Recruitment, growth and moetality of tree species in a lowland dipterocarp forest in Peninsular Malaysia. cited by Pascal and Pelissier, 1996, J. Trop. Ecol. 3, 315–330.

MARD, 2015. Decision 3135/QĐ-BNN-TCLN on forest status. Retrieved from: http://kiemlamvung3.org.vn/quyet-dinh/ (accessed 5.2015).

Marielos, P.C., 2001. Secondary Forest Succession- Processes Affecting the Regeneration of Bolivian Tree Species. PROMAB Scientific Series 3.

MoNRE (Ministry of Natural Resources and Environment of Vietnam), 2011. National report on Biodiversity Year 2011. Ha Noi.

Mueller-Dombois, D., Ellengerg, H., 1974. Aims and methods of vegetation ecology. Wiley, New York.

Nicholson, D.I., 1965. A review of natural regeneration in the dipterocarp forests of Sabah. Malay. Forester 28, 4–25.

Odum, E.P., 1971. Fundamentals of Ecology. W.B. Sanders Co, Philadenphia.

Padalia, H., Chauhan, N., Porwal, M.C., Roy, P.S., 2004. Phytosociological observations on tree species diversity of Andaman Islands, India. Curr. Sci. 87, 799–806.

Pascal, J.P., Pelissier, R., 1996. Structure and floristic composition of a tropical evergreen forest in south-west India. J. Trop. Ecol. 12, 191–214.

Phillips, E.A., 1959. Methods of Vegetation Study. Henri Holt Co. Inc, New York.

Phuc, T.N., Tri, M.T., Son, V.V., 2012. Planting production forest in households in Nam Dong district, model of effectiveness and essentials. Journal of Science, Hue University, Vietnam. Session 72B, No. 3.

PKKNP, 2009. Floristic Composition and Forest Stand Structure in Relation to Topographic Variation in PhuKhaoKhouay National Park in Lao PDR. http://www.sud-expert-plantes.ird.fr/.../797_355_Rapport_Scientifique_dec_2009.doc (accessed 6.2015).

Proctor, J., Anderson, J.M., Chai, P., Vallack, H.W., 1983. Ecological studies in four contrasting lowland rain forests in Gunung Mulu National Park, Sarawak: I. Forest environment, structure and floristics. J. Ecol. 71, 237–260.

Riswan, S., 1987. Structure and floristic composition of a mixed dipterocarp forest at Lampake, East Kalimantan. In: Kostermans, A.J.H. (Ed.), Proc of 3nd Round Table Conference on Dipterocarps, Samarinda, 16–20 April 1985. UNESCO, Jakarta, pp. 437–457.

Rundel, P.W., 1999. Forest Habitat and Flora in Lao PDR, Cambodia and Vietnam. In: Conservation Priorities in Indochina. World Wildlife Fund, Indochina ProgrammeOffice, Hanoi, Vietnam.

Slik, J.W.F., Kessler, P.J.A., van Welzen, P.C., 2004. Macaranga and Mallotus species (Euphorbiaceae) as indicators for disturbance in the mixed lowland dipterocarp forest of East Kalimantan (Indonesia). Ecol. Indic. 2, 311–324.

Sterling, S.J., Hurley, M.M., Minh, L.D., 2006. Vietnam: a natural history. Yale University Press, New Haven, Connecticut (USA).

Thin, N.N., 1994. Diversity of the Cuc Phuong flora – A primary forest of Vietnam. Report of the National Centre of Natural Sciences and Technology, Vietnam 6 (2), 77–82.

Toledo-Aceves, T., et al., 2009. Regeneration of commercial tree species in a logged forest in the Selva Maya, Mexico. For. Ecol. Manage. 258, 2481–2489. cited in Dung, 2010.

Tordoff, A., Timmins, R., Smith, R., Vinh, M.K., 2003. A Biological Assessment of the Central Truong Son Landscape. Central Truong Son Initiative. Report No. 1, WWF Indochina, Hanoi, Viet Nam.

Whitten, A.J., 1982. A numerical analysis of tropical rain forest, using floristic and structural data, and its application to an analysis of gibbin ranging behaviour. J. Ecol. 70, 249–271.

WWF, Saola. http://wwf.panda.org/about_our_earth/species/profiles/mammals/saola/ (accessed 6.15).

6

Simulation of Soil Erosion Risk in the Upstream Area of Bo River Watershed

T.T. Phuong*, R.P. Shrestha†, H.V. Chuong*

*University of Agriculture and Forestry, Hue University, Hue City, Vietnam †Asian Institute of Technology, Pathumthani, Thailand

6.1 INTRODUCTION

Land is degrading and changing significantly under impacts of natural factors and human's resource exploitation. These adverse impacts of natural factors and humans combined with global climate change have caused surface soil to be changed and degraded in quality, especially the status of soil erosion and land degradation in slope soils (Nguyen et al., 2007). Vietnam's statistical data indicate that the total erosion-risk areas are 13 millions of hectares accounting for 40% of natural areas (Nguyen, 2010). The average arable land per capita is decreasing yearly from 0.101 to 0.036 ha per capita. The fragmentation of agricultural land and inappropriate farming techniques lead to low crop productivity and poverty in mountainous regions (Du, 2011). The small watershed in Thua Thien Hue Province is in high risk of degradation, especially erosion (Le and Pham, 2012). In the last 10 years, most of the forestland in the watershed has been changed into other land-use types such as agricultural and residential land. These changes have deep impacts on vegetation cover, surface run-off speed, and hence lead to an increasing trend of soil loss in the watershed (Tran and Huynh, 2013).

Soil erosion causes loss of fertile topsoil cover, and delivers millions of tons of sediment into reservoirs and lakes, resulting in strong environmental impact and high economic costs by its effect on agricultural production, infrastructure, and water quality (Lal, 1998; Pimentel et al., 1995). Not surprisingly soil erosion and sediment delivery have become important topics on the agenda of local and national policy makers. This has led to an increasing demand for watershed or regional-scale soil erosion models to delineate target zones in which conservation measures are likely to be the most effective.

SWAT has been used extensively over the last 30 years all over the world, including in Asia, and has proven to be an effective tool for assessing water resource and nonpoint-source pollution problems for a wide range of scales and environmental conditions (Gassman et al., 2007). Many of the recent studies showed that SWAT is capable to simulate flow, water quality, and soil erosion in large areas, even with limited data, which is important for modelers in developing countries: Mekonnen et al. (2009) applied SWAT to simulate hydrological regimes in two Ethiopian catchments. Quyang et al. (2010) investigated soil erosion dynamics in the upper watershed of the Yellow River, China. The Mekong River Commission has been using SWAT since 2000 to facilitate the joint planning and management of the Mekong River Basin as reported by Rossi et al. (2009). Although SWAT was applied in many other large areas all over the world, the quantity and quality of the spatially variable model inputs as well as the flow and sediment monitoring data for the small watershed of Central Vietnam create a unique opportunity in which we can derive useful experience and techniques to be possibly applied to other small watersheds in all of Vietnam.

Therefore, the goal of this study is to develop a SWAT version 2009 (SWAT, 2009) model (Neitsch et al., 2009) of the small watershed in Central Vietnam, which comprises the upper part of the Bo River watershed. This study was also an additional test case for the efficacy of the SWAT2009 model to represent and simulate spatially variable watershed processes on a small-scale watershed in developing countries where data reliability is often a big issue.

6.2 METHODOLOGY

6.2.1 Study Site Selection

The selected watershed of this study is the upstream area of the Bo River watershed located in the mountainous region of Thua Thien Hue Province, Central Vietnam, with a total area of 14,047.60 ha. The main branches of this river originate from a mountain region with the height of 636 m in the southeast of Luoi District. The upper part of the Bo River goes through the communes of Huong Lam, Huong Phong, and Hong Thuong and is also the basis for boundaries of communes such as Hong Thuong and Phu Vinh communes, Hong Thai and Hong Thuong communes, and Hong Thai and Nham communes. The final point of this river converges on the Sekon River, part of which goes through The People's Democratic Republic of Laos. The watershed has abundant flow dividing the whole basin, and has rainfall intensity, which occurs in steep terrain and cover of mountain ranges. Therefore, the risk of erosion and landslide of riverbanks is very high. The study site is divided into 12 subwatersheds. Most of areas in the study site are natural forest, plantation forest, annual cropland, perennial cropland, and a part of paddy land (Fig. 6.1).

6.2.2 SWAT Model Input

The ArcSWAT2009 model was used for the study to assess soil erosion. SWAT is a basin-scale, continuous-time hydrology model that can produce simulation results on a daily, monthly, or annual basis (Arnold and Fohrer, 2005). The model can simulate stream flow as well as sediment yield. The SWAT model inputs were a digital elevation model

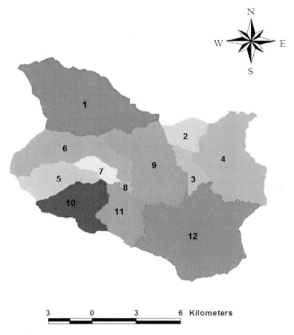

FIG. 6.1   Subwatershed of the study.

(DEM) and land-use maps from 2010 that were obtained from the Department of Natural Resources and Environment (DONRE); a soil map from the Department of Science and Technology that belongs to Thua Thien Hue Province; and weather and hydrology data that were collected from the Hydro Meteorological Center in Hue City and the Institute of Geography, Vietnam.

6.2.3 Watershed Configuration

SWAT divides a watershed into subwatersheds and the subwatersheds can be further subdivided into hydrologic response units (HRUs).

Within each subwatershed, HRUs are formed as unique soil and land-use combinations that are not necessarily contiguous land parcels. In this study, the ArcGIS interface (Winchell et al., 2010) of the SWAT2009 version was used to describe a watershed and extract the SWAT model input files. The DEM was used to delineate the watershed and provide topographic parameters for each subwatershed. The watershed was delineated and described into 12 subwatersheds.

6.2.4 Model Calibration and Validation

The SWAT calibration method was used for the study, which included calibration of the model manually by adjusting hydrologic and sediment parameters in SWAT. The calibration process was basically trial-and-error to yield the highest Nash-Sutcliffe

TABLE 6.1   Calibrated Parameters of the Model

No	Parameters	Range	Calibrated Value
1	Cn2	35-98	89.00
2	Esco	0-1	0.20
3	Surlag	1-24	6.00
4	Alpha_Bf	0-1	0.48
5	Sol_Awc	0-1	0.28
6	Gw_Delay	0-100	31.00

coefficient. Validation is taken to mean "model testing" and validating the model is not necessarily a perfect predictor. Rather, good validation results are simply stronger evidence that the calibrated model is a good simulator of the measured data and does not overmeasure data in the calibration period. In this study, the model was calibrated and validated only for flow due to lack of data on annual sediment load in the outlet station. Flow monitoring data in 2000 and 2010 were used for calibration and validation (Table 6.1).

The coefficient of determination (R^2) and Nash-Suttcliffe coefficient (Nash and Suttcliffe, 1970) were used to quantitatively assess the ability of the model to replicate temporal trends in measured data. The percent bias is defined as the relative percentage difference between time steps.

6.2.5 Soil Erosion Assessment Using SWAT

Erosion and sediment yield in SWAT are estimated of each HRU with the Modified Universal Soil Loss Equation (MUSLE) developed by Wischmeier and Smith (1965; 1978). While the Universal Soil Loss Equation (USLE) uses rainfall as an indicator of erosive energy, MUSLE uses the amount of runoff to simulate erosion and sediment yield. The hydrology mode supplies estimates of runoff volume and peak runoff rate, which, with the subbasin area, are used to calculate the runoff erosive energy. The crop management factor is recalculated every day that runoff occurs. It is a function of aboveground biomass, residue on the soil surface, and the minimum C factor for the plant. The MUSLE is given by

$$\text{sed} = 11.8 \left(Q_{\text{surf}} \cdot q_{\text{peak}} \cdot \text{area}_{\text{hru}} \right)^{0.56} \cdot K_{\text{USLE}} \cdot C_{\text{USLE}} \cdot P_{\text{USLE}} \cdot LS_{\text{USLE}} \cdot \text{CFRG}$$

where sed is the sediment yield on a given day (metric tons); Q is the surface run off volume (mm $H_2O\,ha^{-1}$); q_{peak} is the peak runoff rate ($m^3 s^{-1}$); area_{hru} is the area of the HRU (ha); K_{USLE} is the USLE soil erodibility factor [0.013 metric ton $m^2h/(m^3$-metric ton cm)]; C_{USLE} is the USLE cover and management factor; P_{USLE} is the USLE support practice factor; LS_{USLE} is the USLE topographic factor; and CFRG is the coarse fragment factor.

6.3 RESULTS AND DISCUSSION

6.3.1 The Soil Loss of the Whole Watershed

Output data on total eroded soil for the whole watershed are simulated monthly per year and presented in the file output.std/Annual summary for watershed in the year of simulation.

SOIL LOSS (ton ha^{-1}): Sediment from the subbasin that is transported into the reach during the time step;
PREC (mm): Water that percolates past the root zone during the time step;
SURQ (mm H$_2$O): Surface runoff contribution to stream flow during time step.

As can be seen in Fig. 6.2, the amount of soil erosion over months varies and most of them have reached a peak in October, annually. The correlation between graphs shows that the amount of soil erosion at some point follows the rules of change in rainfall and water flow.

Table 6.2 reveals that the largest amount of soil erosion was 92.33 ton and the highest amount of rainfall was 5624.0 mm in 2007, followed by those in 2005 and 2010, with the highest amount of rainfall and surface runoff being 2427.26 and 2475.44 mm, respectively. In later years, soil erosion tended to change in the amount of rainfall and surface runoff. The average amount of soil erosion was 62.50 ton ha^{-1}, which causes high risk and severely threatens land resources of the whole watershed.

Table 6.3 also shows that the largest amount of soil erosion was 92.33 ton ha^{-1} in 2007, followed by 2010 (85.41 ton ha^{-1}) and 2005 (76.79 ton ha^{-1}). The average amount of soil erosion from 2000 to 2010 was 62.50 ton ha^{-1}, which causes considerably high risk in threatening land resources of the whole watershed (Fig. 6.3).

6.3.2 Soil Erosion at the Subwatershed Level

The amount of soil erosion for each subwatershed is illustrated in the file of output.sub/SYLD (ton ha^{-1}). Fig. 6.4 shows that there is a fair correlation between soil erosion and surface

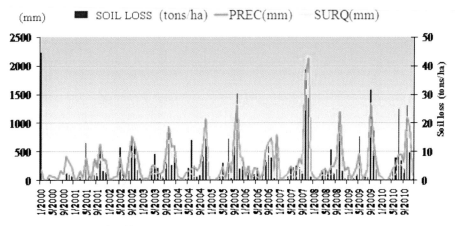

FIG. 6.2 Monthly amount of soil loss.

TABLE 6.2   Soil Loss, Rainfall, and Surface Runoff From 2000 to 2010

Year	Rainfall (mm)	Surface Run-Off (mm)	Soil Loss (ton ha^{-1})
2000	1693.40	383.62	52.09
2001	2725.40	1296.70	37.84
2002	3154.10	1730.04	50.44
2003	3585.70	2156.37	59.73
2004	2988.00	1806.48	55.58
2005	3788.70	2427.26	76.79
2006	3672.70	2305.23	50.31
2007	5624.00	4053.82	92.33
2008	3353.80	1893.16	51.50
2009	3828.70	2421.49	75.48
2010	3961.90	2475.44	85.41
Average	3488.764	2086.328	62.50

TABLE 6.3   Area of Soil Types

Soil Type	Area (ha)	Percent
Ferralic Acrisols_a	8221.85	49.82
Ferralic Acrisols_j	2229.92	13.51
Humic Ferralsols	901.37	5.46
Ferralic Acrisols_s	5150.43	31.21

runoff in the period from 2000 to 2010. However, there still has been a difference among them at some point in time such as 2001, 2008, and 2010. The causes for that are the differences of subwatershed in soil type, land use, and slope. These factors also have a decisive influence on the amount of soil loss due to erosion (Fig. 6.5).

The soil loss of each subbasin varied and there is a deviation between the amount of soil erosion and runoff in the beginning period. Significant correlation between them shows that the amount of soil erosion has tended to vary with surface runoff since 2001. The amount of soil erosion of Sub 8 is the maximum of 19.064 ton ha^{-1} in 2000; while this largest figure of Sub 12 is 153.479 ton ha^{-1} in 2010, which is eightfold in comparison with Sub 8. So it is noted that there is a large difference in the amount of soil erosion because of differences in features of each subbasin in terms of area, width, height, soil type, land-use type, and slope (Table 6.4).

Subwatershed 12 has the maximum average amount of soil erosion with 103.02 ton ha^{-1} year^{-1}, and the minimum is subwatershed 8 with 8.77 ton ha^{-1} year^{-1}. The change in monthly soil erosion of the two subwatersheds is displayed in Fig. 6.6A and B.

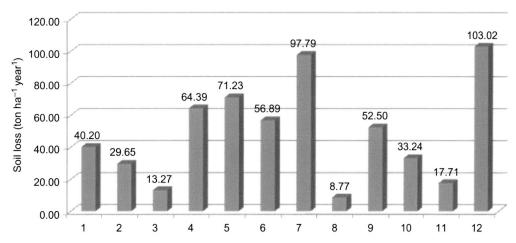

FIG. 6.3 Average soil loss at the subwatershed level from 2000 to 2010.

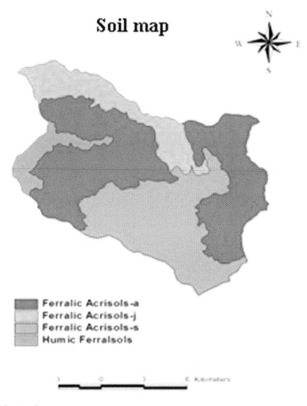

Soil map

Ferralic Acrisols-a
Ferralic Acrisols-j
Ferralic Acrisols-s
Humic Ferralsols

FIG. 6.4 Soil map of the study area.

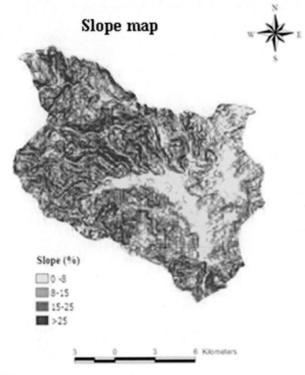

FIG. 6.5   Slope map of the study area.

TABLE 6.4   Area of Slope Land

Slope (%)	Area (ha)	Percent
0-8	739.32	4.48
8-15	1310.34	7.94
15-25	2420.23	14.66
>25	12,033.67	72.92

Table 6.5 shows that soil loss occurs mainly in dry agriculture land area with slope above 25% in Ferralic Acrisols (Fa), while there is a very low amount of soil loss in slope from 8% to 15% with land-use type of forest mixed in Ferralic Acrisols (Fs).

6.3.3 Soil Loss at Different Land-Use Types

According to Neitsch et al. (2009), the canopy affects erosion by reducing the effective rainfall energy of intercepted raindrops. Water drops falling from the canopy may regain appreciable velocity but it will be less than the terminal velocity of free-falling raindrops. The average fall height of drops from the canopy and the density of the canopy will determine the

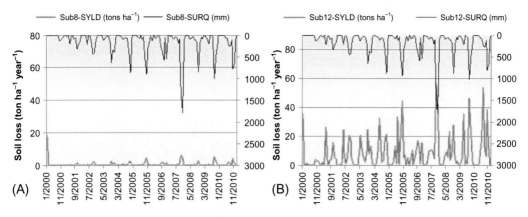

FIG. 6.6   (A) Monthly soil loss from 2000 to 2010 in subwatershed 8 and (B) subwatershed 12.

TABLE 6.5   Land Use, Soil Type, Slope, and Soil Loss in Subwatersheds 8 and 12

Subwatershed	Land-Use Type	Soil Type in FAO-UNESSCO	Slope (%)	Soil Loss (ton ha⁻¹)
8	Forest mixed	Ferralic Acrisols (Fs)	8-15	8.77
12	Dry Agriculture land	Ferralic Acrisols (Fa)	>25	103.02

TABLE 6.6   The Estimate of Soil Loss at Different Land Uses in 2000 and 2010

Land-Use Type	Soil Loss (ton ha⁻¹)		Change (+/−)
	2000	2010	
Wet agriculture land	617.59	267.90	−349.69
Bare land	388.00	844.62	+456.62
Water body	0	0	0
Dry agriculture land	2631.08	24,667.44	+22,036.36
Forest mixed	1540.75	1099.58	−441.17
Residential land	3644.52	2060.58	−1583.94

reduction in rainfall energy expended at the soil surface. The soil loss areas at different land use types in 2000 and 2010 are shown in Table 6.6.

Table 6.6 reveals that the plants with small canopy such as cassava, particularly in dry agricultural land, could be strongly affected by erosion risk. Total soil loss in 2010 of this land-use type increased by 837.54% in comparison with 2000. Meanwhile, wet agriculture land was considered as low soil erosion risk. The largest area in the watershed is forest mixed (Table 6.7 and Fig. 6.7B). However, they are distributed in high slope areas, and a

TABLE 6.7   Area of Land-Use Types in the Study Site

Land Use Type	Area (ha)	Percent
Bare land	80.16	0.49
Dry agriculture land	295.66	1.79
Forest-mixed	14855.54	90.01
Residential land	433.43	2.63
Wet agriculture land	706.35	4.28
Water body	132.43	0.80

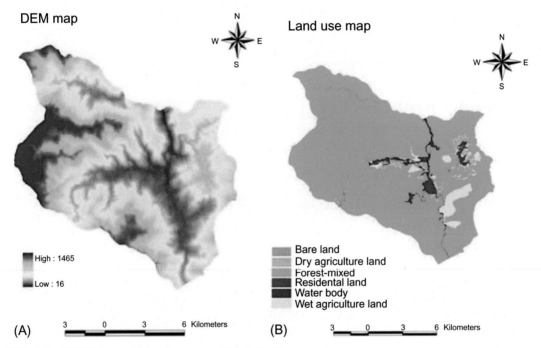

FIG. 6.7   (A) DEM map. (B) Land-use map of the study site.

part of natural forest has changed into plantation forest having trees with small canopies of leaves such as rubber, acacia, and pine. Therefore, soil loss of forest mixed land-use type in 2000 is quite high compared to 2010 with 1540.75 ton ha^{-1}. This is a good sign, thanks to efforts of local authorities and people in protecting forest and greening barren hill under the government's guidelines and policies. The soil erosion risk was decreased 1583.94 ton in 2010 compared to 2000 because this type of land use is increasingly being covered by buildings. The water bodies were neglected because the area of water bodies was not considered in soil erosion risk assessment.

6.3.4 Model Evaluation

Data on sediment measuring in the study site cannot be collected, so in this research runoff data is used to calibrate the model. The process of analyzing sensitivity of runoff is done automatically by the SWAT model in the *"Sensitivity Analysis"* function of ArcSWAT. The results of this process show parameters such as Surlag (the surface runoff lag time), Cn2 (infiltration factor), Esco (the soil evaporation compensation factor), Alpha_Bf (the Alpha factor on baseflow), Sol_Awc (available water capacity of the soil layer), Gw_Delay (groundwater delay) having strong influence on the changing value of runoff volume in rivers of the watershed. Based on the results of this process, the parameters are selected for a process of calibration so that coefficients of evaluation satisfy requirements and obtain the highest accuracy.

In this study, the flow data was applied during the calibration process due to the lack of sediment data. According to Moriasi et al. (2007), model simulation judged as satisfactory if NSE > 0.5 and PBIAS = ±25% for flow (Table 6.8). Therefore, the calibration and validation results of this study can be accepted. The results of the daily flow calibration processes showed good fit between simulated and observed data.

6.3.5 Soil Erosion Risk Mapping

After the erosion database is sorted and formatted appropriately, a soil erosion map is simulated by using the ArcGIS software and adding a field of erosion results of sediment yield (SYLD) (ton ha^{-1}) into an attribute table of HRUs database layer. A soil erosion map of the whole watershed is displayed in Fig. 6.2. Continuously using calculation functions in ArcGIS 9.3 software, the respective area of each erosion level is counted and illustrated in Table 6.9.

Results of erosion risk assessment in the study site by using SWAT and decentralized limitation according to the Land Degradation Inventory Framework of the Ministry of Resources and Environment in 2012 indicate that the majority of the area in the watershed (more than

TABLE 6.8 Model Evaluation Values for Simulated and Observed Stream Flow

Periods	Mean Flow (m^3 s^{-1})		NSE	PBIAS (%)
	Observed	Simulated		
Calibration	25.37	27.42	0.87	3.79
Validation	24.43	28.61	0.65	−8.46

TABLE 6.9 The Area of Soil Erosion Risk in the Study Site

Year	Soil Erosion Risk					
	Low	%	Moderate	%	High	%
2000	2129.76	15.16	7134.30	50.79	4783.54	34.05
2010	535.81	3.81	8512.18	60.60	4999.62	35.59

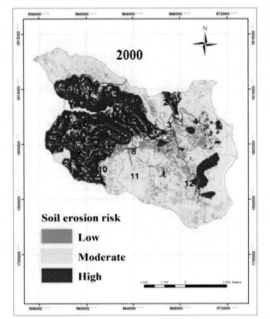 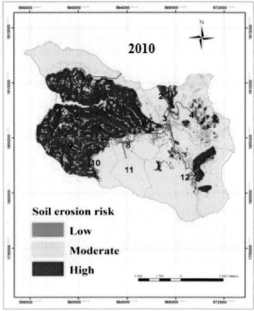

FIG. 6.8   Soil erosion risk in 2000 and 2010.

50%) in 2000 is at a moderate erosion risk level. Especially, high soil erosion risk in the study site still occupies a high percentage in 2000 and 2010, with more than 30% and this trend tends to increase mainly southwest and north of the watershed. The reason is that these areas are often influenced by surface runoff speed, which strengthens the process of separating soil particles and, hence, strengthens erosion (Fig. 6.8).

6.4 CONCLUSION

The results found that the largest amount of soil erosion was 92.33 ton ha^{-1} in 2007, followed by 2010 (85.41 ton ha^{-1}) and 2005 (76.79 ton ha^{-1}). The average amount of soil erosion from 2000 to 2010 was 62.50 ton ha^{-1}, which causes high risk of considerably threatening land resources of the whole watershed. Additionally, the resultant maps of annual soil erosion show a maximum soil loss of 153.48 ton ha^{-1} year^{-1} with a close relation to dry agriculture land areas on slopes higher than 25%. Based on the Land Degradation Inventory Framework of the Ministry of Resources and Environment in 2012, soil erosion risk in the study site was classified. There was a greater than 30% risk of high erosion in the period from 2000 to 2010, and the area at the moderate erosion risk level increased in 2010 compared to 2000. While the area at the low erosion risk level in 2010 had a significant decrease compared to 2000.

The results of the calibration process showed good fit between simulated and observed data; therefore, SWAT can be used to simulate soil erosion over time in the upstream area of the Bo River watershed, a small mountainous watershed of Central Vietnam. Therefore, the research findings of applying this tool will better support resources managers, especially land policy makers and stakeholders involved in the process of land use in the study site in making decisions of land use reasonably and effectively to reduce the effects of soil erosion. However, the application of SWAT required a considerable amount of detailed data, which is not readily available and also it was rather difficult to evaluate data accuracy/reliability because of a certain amount of data.

References

Arnold, J.G., Fohrer, N., 2005. SWAT2000: Current capabilities and research opportunities in applied watershed modeling. Hydrol. Process. 19 (3), 563–572.

Du, L.V., 2011. Farming practices and soil quality. In: International Workshop on Vegetable Agroforestry and Cashew-Cacao Systems in Vietnam. vol. 6a. WASWAC, Ho Chi Minh City, Vietnam, pp. 63–70.

Gassman, P.W., Reyes, M.R., Green, C.H., Arnold, J.G., 2007. The soil and water assessment tool: historical development, applications, and future research directions. Trans. ASABE 50 (4), 1211–1250. American Society of Agricultural and Biological Engineers.

Lal, R., 1998. Soil erosion impact on agronomic productivity and environment quality. Crit. Rev. Plant Sci. 17 (3), 319–464.

Le, P.C.L., Pham, T.T.M., 2012. Assessing potential of land degradation in Thua Thien Hue Province. J. Sci. Hue Univ. 74A (5), 77–84 (In Vietnamese).

Mekonnen, M.A., Worman, A., Dargahi, B., Gebeyehu, A., 2009. Hydrological modelling of Ethiopian catchments using limited data. Hydrol. Process. 23 (23), 3401–3408.

Moriasi, D.N., Arnold, J.G., Liew, M.W.V., Bingner, R.L., Harmel, R.D., Veith, T.L., 2007. Model Evaluation Guidelines for Systematic Quantification of Accuracy in Watershed Simulations, T. ASABE 50, 885–900.

Nash, J.E., Suttcliffe, J.V., 1970. River flow forecasting through conceptual models, Part I. A discussion of principles. J. Hydrol. 10 (3), 282–290.

Neitsch, S.L., Arnold, J.G., Kiniry, J.R., Srinivasan, R., Williams, J.R., 2009. Soil and Water Assessment Tool, Theoretical Documentation: Version 2009. USDA Agricultural Research Service and Texas A&M Blackland Research Center, Temple, TX.

Nguyen, A.H., 2010. Comprehensively Studying Geographical Arising and Land Degradation Aiming the Purpose of Reasonably Using Land Resource and Preventing Disaster in Binh-Tri-Thien region. Doctor of Philosophy dissertation. Hanoi., (In Vietnamese).

Nguyen, D.K., et al., 2007. Studying, assessing and forecasting land degradation in the North Central of Vietnam aiming sustainable planning. Final report of research run by Ministry of Science and Technology. Hanoi, (In Vietnamese).

Pimentel, D., Harvey, C., Resosudarmo, P., Sinclair, K., Kurz, D., McNair, M., Crist, S., Shpritz, L., Fitton, L., Saffouri, R., Blair, R., 1995. Environmental and economic costs of soil erosion and conservation benefits. Science 267, 1117–1123.

Quyang, W., Skidmore, A.K., Hao, F., Wang, T., 2010. Soil erosion dynamics response to landscape pattern. Sci. Total Environ. 87 (6), 1358–1366.

Rossi, C.G., Srinivasan, R., Jirayoot, K., Le Duc, T., Souvannabouth, P., Binh, N.D., Gassman, P.W., 2009. Hydrologic evaluation of the lower Mekong River basin with the Soil and Water Assessment Tool model. Int. Agric. Eng. J. 18 (1–2), 1–13.

Tran, T.P., Huynh, V.C., 2013. Simulating effects of land use change on soil erosion in Bo River basin in the Central of Vietnam. J. Agric. Rural Dev. 2 (In Vietnamese).

Winchell, M., Srinivasan, R., Di Luzio, M., Arnold, J., 2010. Arc SWAT interface for SWAT 2009. Users' guide. Grassland, Soil and Water Research Laboratory, Agricultural Research Service, and Blackland Research Center. Texas Agricultural Experiment Station: Temple, Texas 76502, USA, p. 495.

Wischmeier, W.H., Smith, D.D., 1965. Predicting rainfall-erosion losses from crop land east of Rocky Mountain. Agr. Handbook No. 282, U. S. Dept. Agr., Washington, DC.

Wischmeier, W.H., Smith, D.D., 1978. Predicting rainfall erosion losses from crop land east of Rocky Mountain. Agr. Handbook No. 537, U. S. Dept. Agr., Science and Education Administration.

MERGING SCIENCE AND TRADITIONAL PRACTICES IN NATURAL RESOURCE MANAGEMENT

7

Propagation of *Scaphium lychnophorum* Pierre. and *Baccaurea sylvestris* Lour. for Enriching Community Forests

H.T.H. Que, T.N. Thang

Hue University of Agriculture and Forestry, Hue, Vietnam

7.1 INTRODUCTION

Uoi bay (*Scaphium lychnophorum*) and *Dau sac* (*Baccaurea sylvestris*) are endemic forest trees that provide fruits of high economic value. Fruits of *Uoi bay* provide medicines and drinks, which are a large income source for forest-dependent households in Thua Thien Hue Province (Hy, 2005). *B. sylvestris* provides edible and tasteful fruits for the local market. *Uoi bay* was listed in the International Union for Conservation of Nature (IUCN) Red List as "least concern" (IUCN, 2014), but this species is facing a high risk of extinction because the mother trees were often cut during the process of fruit harvesting.

Since the 1990s, the government of Vietnam has started forestland allocation policy that encouraged local people to participate in forest management through sustainable uses of forest products and the enhancement of forest enrichment and rehabilitation. The allocation policy provided forestland tenure to households or communities in 50 years and can be renewed provided that forestland users adopt sustainable land uses. According to the Law on Forest Protection and Development (2004), the forestland users must conserve natural forests and can only grow plantation on bare land. Because most allocated forests were degraded, the local people often harvest nontimber forest products (NTFPs) to either supply their food demand or cover costs for forest management.

The practice of cutting endemic tree species such as *Uoi bay* and *Dau sac* has negatively influenced forest diversity of allocated forests. In addition, this cutting could easily lead to the larger deforestation to have land available for planting exotic species of acacia, which can provide faster income. Therefore, propagation of *S. lychnophorum* and *B. sylvestris* tree species has an important meaning in both conserving endemic tree species in the wild and providing sufficient amount of fruits for income generation. Through propagation, we expected to provide seedlings for NTFPs enrichment in natural forest, and shorten fruiting time, so that farmers can grow these trees in their allocated forests for both conservation and livelihood purposes.

7.2 METHODOLOGY

The study was conducted in Nam Dong District, which is located in the southwestern part of Thua Thien Hue Province, Vietnam. This district has recorded 83 species of NTFPs that were collected from natural forests (Wetterwald et al., 2004). Secondary data were collected on different techniques of propagation, including seed germination (seedlings), grafting, and cutting. After reviewing propagation techniques, we decided to apply grafting technique for *Uoi bay* and seedlings for *Dau sac*.

7.2.1 Concept of Seed Propagation

Many tree species can be propagated by seeds where the soil and climatic conditions are suitable. But if the seeds are from the same source that is sown at different locations, it will give different results of growth, flowering, and fruiting of plants (Kumar, 1999). Propagation from seed is less expensive. The seed is kept dry and cool for maintaining its viability from harvest to the next planting season. Many seeds can be stored for many years under suitable conditions. Propagation by seed controls virus diseases, as virus diseases are not transmitted by the seed (Kumar, 1999). Propagation by seeds is the major method by which plants reproduce in nature and one of the most efficient and widely used propagation methods for cultivated crops. The plants produced are referred to as seedlings (Hartmann et al., 2002). Propagation by seeds also provides propagation materials in the form of cuttings and grafts. However, seed propagation has two disadvantages: (1) genetic variation occurs in seed from cross-pollinated plants and therefore the plant grown from seed may not exactly duplicate the characteristics of its parents and many undesirable characteristics may be seen; and (2) some plants take a longer time to grow from seed to maturity.

7.2.2 Concept of Grafting Propagation

Grafting is the art of connecting two pieces of living plant tissue together in such a manner that they will unite and subsequently grow and develop as one composite plant. Fruit and nut trees and woody plant species have been grafted because of difficulty in propagation by cutting, and grafted trees yield higher value.

Grafted and budded plants can improve plant quality, fruit yield, superior forms, and greater plant ecological ranges. Grafting and budding also serves diverse purposes such as uniformity of populations, obtaining benefits of certain rootstocks and scion, shortened flowering time, and elimination of virus diseases (Hartmann et al., 2002). For successful grafting, Hartmann et al. (2002) described five important elements: (1) The rootstock and scion must be compatible; (2) the vascular cambium of the scion must be placed in intimate contact with that of the rootstock; (3) the grafting operation must be done at the time when the rootstock and scion are in the proper physiological stage; (4) immediately after the grafting operation is completed, all the cut surfaces must be protected from desiccation; and (5) proper care must be given to the grafts for a certain period after grafting.

7.2.3 Data Collection

Primary data were collected in different steps depending on particular techniques of propagation. For the seed propagation (*Dau sac*), six steps were undertaken: (1) viability testing to determine the number of live seeds in the sample, (2) conducting warm water pretreatment, (3) germination testing, (4) forming a sampling design, (5) forming a block design, and (6) undergoing tending method. For the graft propagation (*Uoi bay*), there were four steps undertaken: (1) grafting (selection of compatible rootstock and scion), (2) formulating a sampling design, (3) forming a block design, and (4) tending method.

In the nursery experiment, the average temperature recorded was 26.2°C (lowest 21.3°C and highest 35.6°C). Rainfall from late August to early of October (about 2 months) was approximately 820.5 mm, and the average humidity was 88%.

The data was quantitatively analyzed through descriptive and analytical statistics especially analysis of variance (ANOVA) through the Statistical Package for the Social Sciences (SPPS). Germination percentage, viable and emerged bud percentage were analyzed accordingly.

7.3 RESULTS AND DISCUSSION

7.3.1 Species Description

Dau sac (*B. sylvestris*) is an evergreen wood tree species, 14–15 m in height and 30–50 cm in diameter. Stems are round and have an angle with smaller support. Branches incline as they grow while occasionally crisscrossing. The crown is darker green with pinkish brown bark along with rippled cracked boles. Leaves that are complete are single and alternate, which are crowded at branch heads. Leaves that are oval and sharp are 10–12 cm long, 5–7 cm wide with side-veining antithetic with clear emersion. Leaf stalk is round, which is 18–19 cm in length with two swollen heads. Fruits are multiloculated, 1.7–1.8 cm in diameter; fruit-stalk is very short, and each fruit has six incisions that open three dry valvars; and ripe fruit has a yellow color. Fruit can be edible and has a sweet-sour taste, usually with 1–5 seeds.

Dau sac is an endemic tree that grows in natural forest, and generally grows as a tropical evergreen- broad-leaf forest, which may be either primary or secondary forest. It is distributed in the hilly areas of Nghe An, Ha Tinh, Quang Binh, Quang Tri, Thua Thien Hue, Gia Lai, Kon Tum, Quang Ngai, and Binh Dinh provinces. Trees are medium and are susceptible to shade but require moist, thick layered, moist sloppy terrain with ferritic soil. Trees require enough shadow with 0.3–0.6 cover degrees for regeneration. Flowering occurs in June while fruiting is in September. Wood pith is not discriminated with a pinkish brown soft silky cut surface (Figs. 7.1 and 7.2).

Uoi bay (*S. lychnophorum* or *S. macropodum*) is a huge wooden tree, which is 20–25 m in height. Stems are straight while the bark is stringed, having light yellow tender hair branches. Leaves are single and alternate, often with crowned branch head that are lobate in a young

FIG. 7.1   Bunch of *Baccaurea sylvestris* fruits.

FIG. 7.2   Seeds of *Baccaurea sylvestris*.

tree while the leafstalk is 10–20 cm in length. Flowers are polygamous that produces difollicular fruit with thick skin. Each follicular has one dark brown wrinkled seed. *Uoi bay* requires light that promotes faster growth that tends to flower in March-April and ripens in June-August. It is a pioneer species to promote reforestation in natural forest, especially in central and highlands of Vietnam.

Uoi bay is distributed most in natural forest. Seedlings are regenerated quickly at higher density in the wild (150–200 seedling/ha). It has a fruiting cycle every 2 or 3 years. Wood is soft and has a beautiful vein, which often is used for fiber and making utensils. Fruits are used as medicine or soft drinks. However, often local farmers in Nam Dong and A Luoi districts harvest the fruits through whole tree felling (Hy, 2005).

7.3.2 Seed Propagation of *B. sylvestris*

7.3.2.1 Cutting Test

We cut 100 seeds of *B. sylvestris* to examine the embryo and endosperm by using razor blades. Seeds were cut longitudinally through the embryo. Attempts were made to examine embryo and endosperm. Firm, light green embryo are often considered healthy while the percentage of viable seeds was 100% after 2 weeks of storage.

7.3.2.2 Germination Testing

The germination test result is presented as average cumulative germination percentage and standard deviation (SD) for each treatment, which is presented in Table 7.1.

Germination time was not significantly different for the three treatment methods ($p > 0.05$, ANOVA). Seeds began germinating after 13 days at the same day for the three treatments and continued germinating for 5 days. The rate of germination was quite high. Germination percentage was significantly different among treatment methods ($p < 0.05$, ANOVA). Germination percentage of soaking in hot water was the highest ($99.2 \pm 1.1\%$). The lowest germinated percentage was $81.6 \pm 2.6\%$ for control treatment.

7.3.3 Grafting Propagation of *S. lychnophorum*

The result of propagation is presented as average percentage of viable buds, average cumulative of emergent percentage, average emergent time, and the SD enlisted in Table 7.2.

Percent of viable buds was not significantly different for the three grafting methods ($p > 0.05$, ANOVA), but there were significant differences of emergent percentage and emergent time for the three grafting methods ($p < 0.05$, ANOVA). The rate of emergence of chip budding was the highest ($42.2 \pm 3.85\%$). The lowest emergent percentage was 2.2 for the patch budding

TABLE 7.1 Seed Propagation of *Baccaurea sylvestris*

Treatment Methods	Control (SD) ($n = 50$)	Soaking in Hot Water (70°C) (SD) ($n = 50$)	Soaking in Normal Water (SD) ($n = 50$)	Significance (*p*-Value)
Germination time (day)	14.71 (0.37)	14.45 (0.10)	14.58 (0.11)	>0.05
Percent of germination	81.6 (2.6)	99.2 (1.1)	93.6 (2.6)	<0.05

TABLE 7.2　Results of Grafting Propagation of *Scaphium lychnophorum*

Methods	Chip Budding (SD) ($n=15$)	Side Grafting (SD) ($n=15$)	Patch Budding (SD) ($n=15$)	Significance (p-Value)
Percent of viable buds (%)	68.9 (3.85)	62.2 (3.85)	60.0 (6.67)	>0.05
Emergent time (day)	15.3 (0.30)	11.8 (0.43)	20.0 (0.00)	<0.05
Emergent percentage (%)	42.2 (3.85)	31.1 (3.84)	2.2 (3.85)	<0.05

method. The emergent time of the side grafting method was the shortest (11.8±0.43 days). For the chip budding and side grafting methods, each method had its own strong and weak points. The chip budding method had emerged buds time longer and growth weaker than side grafting but the percentage of emerged buds was higher and reduced the bud-stick quantity that was required. Contrary to chip budding, although the side grafting method had a lower percent of emerged buds and bud-stick quantity that was required was more than chip budding but emerged buds time was shorter and emerged buds grew stronger than other methods. The side grafting method can also make use of soft scion that cannot be used in chip budding. The slowest emergent time was 20 days for patch budding and emerged bud died after that several days.

7.4 CONCLUSIONS AND RECOMMENDATIONS

A cutting test revealed that *B. sylvestris* propagation by seeds could yield 100% viable seeds with germination time not significantly different for the three treatments, but the rate of germination by soaking in hot water 70°C for 24 h was the highest (99.2%) and optimal. Nevertheless, these seedlings are easily infected by collar rot, which may lead to mass death and requires frequent monitoring and medication to be offered accordingly.

Due to unstable physiological rootstocks, propagation of *S. lychnophorum* took longer time as the rootstocks had been transplanted from a local nursery to the research site 1 month before grafting. This was further aggravated by higher mean relative humidity (88%), which lead buds to die easily and form callus quite slowly. However, *S. lychnophorum* tree had good propagation capacity by grafting. The chip budding method had the highest percent of emerged buds while buds through side grafting emerged in the shortest time.

1. *B. sylvestris* seeds require soaking in hot water (70°C) and normal water for 24 h before sowing to give better results.
2. For the propagation of *S. lychnophorum* by grafting, both chip budding and side grafting (branch grafting) yield better results.
3. With improved techniques in propagation of *B. sylvestris* and *S. lychnophorum*, these native species will be conserved while pressure on forest is likely to be reduced.

References

Hartmann, H.T., Kester, D.E., Fred, T.D.J., Robert, L.G., 2002. Plant Propagation: Principles and Practices, sixth ed. Prentice-Hall, New Delhi.

Hy, H., 2005. *Scaphium lychnophorum*. Scientific and Industrial News-Letter in Thua Thien Hue Province 7, 26–27.

IUCN, 2014. The IUCN Red List of Threatened Species. Version 2014-4. http://www.iucnredlist.org.

Kumar, V., 1999. Nursery and Plantation Practices in Forestry, second ed. Scientific Publishers, India.

Wetterwald, O., Zingerli, C., Sorg, J.-P., 2004. Non-timber forest products in Nam Dong District, Central Vietnam: ecological and economic prospects. Schweiz. Z. Forstwes 155 (2), 45–52.

8

Fishery Communities' Perception of Climate Change Effects on Local Livelihoods in Tam Giang Lagoon, Vietnam

H.D. Ha, T.N. Thang

Hue University of Agriculture and Forestry, Hue, Vietnam

8.1 INTRODUCTION

8.1.1 Background

Running parallel with the coast of Thua Thien Hue Province, Tam Giang lagoon receives freshwater resources from main rivers such as O Lau, Bo, Perfumer, Dai Giang, and Truoi (Cu, 1996). This lagoon is also connected with the ocean by Thuan An opening. The interference of water resources from the rivers and the sea makes the lagoons have a special brackish water environment, and creates a high biological multiform and abundant organism resources including over 1000 species, of which there are 150 species that bring high economic values (Thung, 2006).

Residential communities along the lagoon are concentrated where the people live in the units and civil society is diverse and abundant. If taken on professional criteria, cultural or other forms of exploitation and use of lagoon resources to divide the village along lagoon have other forms: (1) the agricultural village, with farming and rural type fishermen; and (2) the village consists of fisheries or fishing village settlements on the rivers, lagoons, and estuaries. Along with the formation of the village, during development the coastal towns and cities are increasing, dominating all aspects contributing to the rapid development of a coastal lagoon (Tuyen et al., 2010).

The advantages of the coastal lagoons and coastal communities include the available aquatic resources with abundant reserves that facilitate the process of exploitation and use for economic and social development (Nguyen, 2008), which is one of the advantages of the coastal areas in comparison with inland areas.

8.1.2 Problem Statement

In recent years, natural hazards are more serious day by day in the Tam Giang area, and studies of climate change have warned of the dangers of climate change's impacts for development and survival of people. The coastal zone is the area that is most vulnerable to climate change. Authorities and coastal communities have been trying to adapt to climate change.

Although many research studies have been applied in the Tam Giang lagoon related to climate change effects, there have not been in-depth studies to analyze the perceptions of fishermen. In the new context of climate change, the assessment perception of fishermen will point out the solutions for fishery communities to copy in light of changing weather and climate, and ways to promote these strengths to adapt to climate change in Tam Giang lagoon.

In addition, the solutions given by the community to adapt to climate change should be further analyzed to indicate their contribution to adaptation and mitigation of impacts of climate change on the living processes of coastal fishing communities.

8.1.3 Objectives of the Study

The overall objective of this research is to assess the perception of fishery communities to the effect of climate change in Tam Giang lagoon.

Specific objectives

1. To describe the manifestations of climate change in the study area.
2. To evaluate the perception of fishery communities about climate change effects.

8.2 METHODOLOGY

8.2.1 Selection of Study Area

This study was conducted in two fishing communities: Quang An Commune, which is located in the north, and Huong Phong Commune, which is located in the middle of the lagoon.

The two sites selected contain specific natural conditions, and social aspects such as livelihood and occupation of the local people and can be representative of the whole lagoon (Map 8.1).

8.2.1.1 Quang An Commune

Quang An Commune is located in the west of Tam Giang lagoon, Quang Dien District, Thua Thien Hue Province. Quang An also located in the middle of the Tam Giang lagoon, about 7 km from Thuan An seaport in the northwest and 12 km from Hue City to the west-southwest. The lagoon area is mainly concentrated in An Xuan and My On hamlets.

An Xuan and My On are two fishing communities where people's livelihoods are closely associated with lagoon natural resources exploitation. The two communities have all the features of the coastal villages of Tam Giang lagoon like major fishery households, a rapid rate of population growth, a young population structure, and abundant labor resources (Quang An, 2011).

Map 8.1 Location of study areas

• Research site: Tam Giang – Cau Hai lagoon, Thua Thien Hue province

MAP 8.1   Location of study areas.

8.2.1.2 Huong Phong Commune

Huong Phong is a commune located in the northwest of Huong Tra District, Thua Thien Hue Province, 12 km from Hue City. With a total natural area of 1569 ha, Huong Phong is located in a special place; both sides of this commune border the river and water surface of Tam Giang lagoon, so Huong Phong has advantages in agriculture and aquaculture development (Map 8.2).

Huong Phong Commune is often influenced by seawater and freshwater in the summer and flooding season, respectively. Huong Phong has 532.08 ha of water surface for fishing and aquaculture (Huong Phong, 2011).

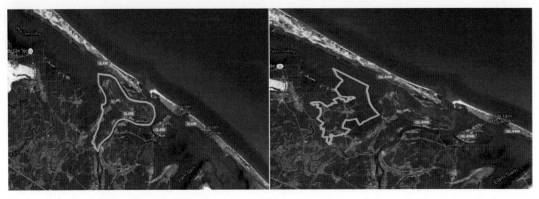

MAP 8.2   Map of Huong Phong and Quang An Communes.

8.2.2 Household Occupation

The criteria for selecting the study areas are:

Household selection criteria: The interviewed households were selected based on the following criteria:

- Household must be a member of fishery communities in the Tam Giang lagoon.
- Their livelihoods are directly related to the water surface of lagoon areas: aquaculture and fishing operations.
- Income from fisheries is a major source of household income.
- There are both poor and nonpoor households (follow the Vietnam poor criterion in 2011).

8.2.2.1 Sample Methods and Sample Size

A random sampling method was used to collect samples for this study. A list of fishery households in Quang An and Huong Phong selected for interviews were randomly selected by the community leaders and research team.

The sample size was calculated by using the formula below and based on the number of fishery households in Quang An and Huong Phong:

$$n = N / \left(1 + N \times e^2\right)$$

In the case of this study, n is the sample size (number of households chosen to interview); N is the total fishery households in study areas; e is the level of error ($N = 451$; $e = 8.5\%$; $n = 105$).

The interviewed households in Quang An totaled 60 households and Huong Phong was 45 households. In general, the workforce of Quang An and Huong Phong is abundant and mainly engaged in the agricultural and fishery areas.

Table 8.1 shows the differences in occupation structure in two study areas. In Quang An, the number of fishing households is 55% more than the number of aquaculture households (33.3%). In Huong Phong, the number of aquacultural households is more than fishing households.

TABLE 8.1   Distribution of Respondents by Major Occupation

Occupation	Quang An, N = 60 (%)	Huong Phong, N = 45 (%)	Total, N = 105 (%)
Fishing	55	40	48.6
Aquaculture	33.3	48.9	40
Agriculture + Aquaculture	11.7	11.1	11.4
Total	100	100	100

Field survey, 2010.

Although Quang An has an advantage in agricultural land area over Huong Phong, the actual results of the investigation show that the percentage of households having both aquaculture and agriculture is equivalent with more than 11%.

8.2.3 Quantitative Analysis

The quantitative data of this research was processed through the Excel 2010 software program, and analyzed by descriptive and statistical analysis to show the results.

8.2.3.1 Weighted Average Index

Weighted average index (WAI) was used to convert the ordinal data into scale data to analyze the level of perception of local fishermen about climate change effects on their livelihood.
The formula of calculation of WAI is as follows:

$$WAI = \sum S_i F_i / N$$

where WAI is the weighted average index ($0 \leq WAI \leq 1$), S_i is the scale value assigned at its priority, F_i is the frequency of household's respondents, and N is the total number of observation.

These indices were designed based on social scale; the value of each index was kept from 0 to 1. The kind of each index is mentioned as follows.

PERCEPTION INDEX

This index is used to evaluate the level of fishermen's perception about climate change effects in study areas. It includes five levels: very strong, strong, medium, weak, and very weak.

Categories	Very Strong	Strong	Medium	Weak	Very Weak
Scale	1	0.75	0.5	0.25	0

SATISFACTION INDEX

The formula for calculating satisfaction index is as follows:

$$WAI = \left(1.00 \times f_1 + 0.75 \times f_2 + 0.5 \times f_3 + 0.25 \times f_4 + 0 \times f_5\right) / N$$

where WAI is the weighted average index, $0 \leq WAI \leq 1$; f_1 is the frequency of first scale choice; f_2 is the frequency of second scale choice; f_3 is the frequency of third scale choice; f_4 is the frequency of fourth scale choice; f_5 is the frequency of fifth scale choice.

8.3 RESULTS AND DISCUSSION

8.3.1 Climate Change Situation in Thua Thien Hue Province

According to the records of the Thua Thien Hue Department of Agriculture and Rural Development in 2010, the trend of natural disasters has been increasing in Thua Thien Hue Province recently. The natural disasters have more adverse effects than in the past, affecting agricultural production and the economic development process of the community in general, especially poor communities whose livelihoods are based on access to natural resources (Thua Thien Hue DARD, 2010).

As discussions with key informants who are knowledgeable in the field of climate change in Thua Thien Hue Province show, Thua Thien Hue has been affected by the impacts of climate change. Natural disasters related to climate change are summarized in Table 8.2. However, the most serious natural disasters, including floods, storms, increased rainfall, sea level rise, and extreme cold, will continue to occur frequently in Thua Thien Hue with severe impacts on local communities.

There are many types of natural disasters related to climate change in Thua Thien Hue Province (Thua Thien Hue CPC, 2010); however, floods, storms, temperature changes, and sea level rise are the most the prevalent ones.

8.3.1.1 Floods

Floods are natural phenomena that occur frequently in Thua Thien Hue Province every year. Synthesis of the results reports by the Center for Hydrometeorology Prediction (CHP) in 2009 showed that there were significant changes in the flood season: the main flood season is usually from late Sep. to late Nov. and the early flood season appears from May to Jun. Report analysis results from Thua Thien Hue CHP show that the unusual appearance of the rainy season and an unseasonably rainy Apr. caused flooding to adversely affect agricultural and aquaculture production (Center Hydrometeorology Prediction, 2009).

TABLE 8.2   Types of Natural Disasters That Affect Thua Thien Hue Province

High Impact	Medium Impact	Low Impact
1. Storm	1. Flash flood	1. Acid rain
2. Flood	2. Landslide	2. Drought
3. Precipitation	3. Coastline erosion	3. Saline intrusion
4. Extreme cold	4. Whirlwind	
5. Sea level rise	5. Monsoon	

Cited to NAV, 2010. Situation Analysis on Climate Change in Thua Thien Hue Province, Vietnam. Nordic Assistant to Vietnam annual report. Hue, Vietnam. Nordic Assistance to Vietnam Project, and Key informant group discussion, 2011.

Number of flood events: According to monitoring data from 1977 to 2006 on the Perfume River, the number of flood in this river basin was 3.5/year averagely and 36% of them were large floods. In the La Niña phenomenon years, flooding and flood peak is significantly greater (Nguyen Viet, 2008).

8.3.1.2 Rainfall

Rainfall has decreased between Jun. and Aug. but increased from Sep. to Nov. At that time of year, floods occur regularly but now they occur with a frequency and intensity more than before 1990. Overall, there was a significant increase in rainfall from the 1970s to the 2000s, from less than 2000 mm per year in 1974 to close to 3000 mm in 2004 (see Fig. 8.1). Between 1978 and 2008 there were 108 floods where the water level reached the second warning level. On average, Thua Thien Hue has about 3.5 floods per year (NAV, 2010) but in 2007 nine floods occurred in provinces during the rainy season.

According to Nguyen and Nguyen (2010), the average annual precipitation increased by 7% in Hue, but during the dry season it reduced from 10% to 15%. In contrast, in the rainy months it increased from 10% to 24%. The decline in rainfall in the dry season usually causes the long-term phenomenon of drought and severe impact on people's living and production activities.

The increases in frequency and intensity of floods have caused annual flooding in lowlands, coastal, and mountainous areas where the poor communities are living in Thua Thien Hue. Map 8.3 shows that the lagoon where this study was conducted is the most severely affected by floods.

The increase in frequency and severity of annual flooding can cause tremendous damage to the local communities. However, in other aspects, floods also have some advantages; for example. silt to farmland, clean polluted water environment on channels, rivers, lagoons, especially in bringing source of income for mobile gear fishery households.

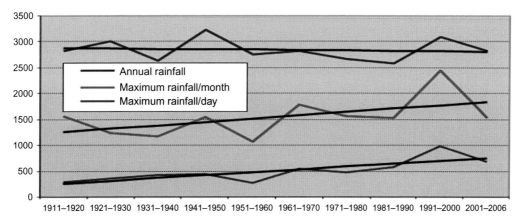

FIG. 8.1 The rainfall variation from 1911 to 2005 in Thua Thien Hue Province. *NAV, 2010. Situation Analysis on Climate Change in Thua Thien Hue Province, Vietnam. Nordic Assistant to Vietnam annual report. Hue, Vietnam. Nordic Assistance to Vietnam Project.*

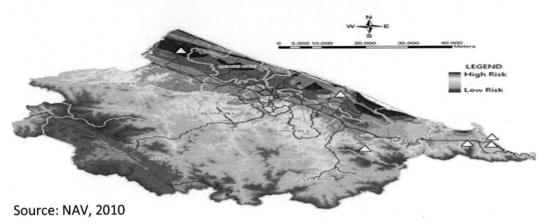

Source: NAV, 2010

MAP 8.3   Elevation map showing flood risks in Thua Thien Hue. *NAV, 2010. Situation Analysis on Climate Change in Thua Thien Hue Province, Vietnam. Nordic Assistant to Vietnam annual report. Hue, Vietnam. Nordic Assistance to Vietnam Project.*

8.3.1.3 Storms

The storm statistical report of CHP shows that storms, tropical depressions, and severe natural disasters occur frequently in Thua Thien Hue. Hurricane season is usually from Aug. to Nov. Nov. is recognized as the month of storms. Forty-one of the hurricanes and 100 tropical storms have influenced Thua Thien Hue Province in the last 50 years.

This study found that there is no change in the number of storms, but the extent of the storms and tropical depressions has increased and the path of the storms has moved to the south (Fig. 8.2). These changes in magnitude and direction make it difficult for government to provide a precise early warning message and response measures for the local community.

Thus, recent hurricanes and tropical depressions have caused more damage to local communities that are located in coastal lagoons and uplands. The increase in the intensity of the storms leads to an increase in risk for the manufacturing sectors, especially agriculture and fishery production.

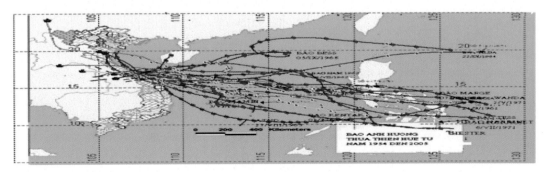

FIG. 8.2   Distribution of storms hitting Thua Thien Hue from 1954 to 2006. *Hydro-metrological Centre, 2009.*

8.3.1.4 Temperature (Extreme Cold/Hot)

The data from CHP shows that the temperature has increased by 1°C in the last 30 years in Thua Thien Hue Province. There was also a significant difference in temperature between winter and summer.

In summer, temperature and very hot days have increased in the last 5 years and caused negative impacts to agriculture, fisheries, and livelihoods of local communities. The increase in temperature affects both coastal communities and the mountainous areas of Thua Thien Hue.

CHP also pointed out that the extremely cold periods tend to be longer than before. Recent studies show that in the 1990s, extreme cold usually lasted for 10–15 days, but in 2007 and 2008, extreme cold lasted about 45–50 days (NAV, 2010). This has a very big impact on aquaculture households in the cold province; aquaculturists must adjust their seasonal schedule to suit the weather conditions.

8.3.1.5 Sea Level Rise

Thua Thien Hue is a coastal province; thus it is threatened by rising sea levels and saltwater intrusion. Over the years, erosion has occurred along the 70 km coastal areas of Thua Thien Hue Province.

In terms of salinity intrusion, the extreme hot summer combined with sea level rise has caused saltwater intrusion in the lagoon and the coastal communes in Thua Thien Hue. Salinity had a negative impact on agricultural production and daily life of the locality.

The sea level rise scenario predicted sea level and saltwater intrusion would increase in severity in the future, affecting the agricultural sector in the districts around the Tam Giang-Cau Hai lagoon (MONRE, 2009). For example, if the sea level rise 22 cm, large areas of rice fields and the low areas and residential areas in Phu Loc, Quang Dien, Huong Tra, and Phu Vang will be flooded and livelihoods of local communities living around the lagoon will be threatened.

However, sea level rise also has a positive impact on water quality and salinity in the Tam Giang-Cau Hai lagoon, such as increasing the aquatic breeding stock source from the ocean and high salinity helps aquaculture production.

8.3.2 Perception of Fishery Communities About Climate Change Effects in Study Areas

Perception of fishermen and aquaculturists (fishery community) living in Tam Giang lagoon on the effects of climate change is considered an important basis for scientists and local government to build adaptation measures to mitigate the impact of climate change on the livelihood and living conditions of fishermen. We select two main groups of local resource users, namely, fishing group and aquaculture group, to evaluate their perception on the influence of climate change on communities through different livelihood types at Tam Giang lagoon.

8.3.2.1 Perception of Fishery Communities About Storm Effect

Table 8.3 shows the results of WAI to assess the level of awareness of local people about the impact of storms to their livelihoods. The average WAI index is 0.76, which means that local people assess storm impact as strong on fisheries production activities of the community at research sites.

Table 8.3 also shows that two groups, fishing and aquaculture in Quang An and Huong Phong, perceive the same level of climate change effect on their livelihood. WAIs ranging

TABLE 8.3   Fishery Community's Perception of Storm Effect on Livelihood

Level of Climate Change Effect	Quang An				Huong Phong				Total	
	AH		FH		AH		FH			
	N=29	%	N=31	%	N=27	%	N=18	%	N=105	%
Very strong	14	48.3	18	58.1	9	33.3	10	55.6	51	48.6
Strong	9	31	4	12.9	8	29.6	4	22.1	25	23.8
Medium	2	6.9	6	19.4	6	22.3	2	11.1	16	15.2
Weak	3	10.3	1	3.2	4	14.8	1	5.6	9	8.6
Very weak	1	3.5	2	6.4	0	0	1	5.6	4	3.8
WAI	0.78		0.78		0.70		0.79		0.76	
OA	S		S		S		S		S	

Note: Very weak (VW): 0.01–0.2; weak (W): 0.21–0.4; medium (M): 0.41–0.6; strong (S): 0.61–0.8; very strong (VS): 0.81–1.
OA, overall assessment; AH, aquaculture household; FH, fishing household.
Field survey, 2011.

from 0.70 to 0.79 mean that the storms caused a decrease in fishery production, destroying infrastructure such as ponds, fishing gear, houses, and roads. In general, storms strongly affect the production activities of fishermen in both Quang An and Huong Phong.

This result is consistent with results reported from 2006 to 2010. The report shows that in Quang An and Huong Phong, storms are considered the most powerful factors that influence production at the local fishery. The annual hurricane caused great loss of life and property in two communes. Details and percentages for two indicators point out the WAI of fishing and aquaculture groups in Table 8.3.

8.3.2.2 Fishery Community's Perception of Flood Incidence on Livelihood

In general, annual flooding affects greatly the fishery production activities in coastal communes in Tam Giang lagoon. The change in number, appearance, time, and intensity of floods caused these effects to varying degrees to the aquaculture production groups.

The level of awareness of local people about the impact of floods is different between groups of fishing and aquaculture in Quang An and Huong Phong. Table 8.4 shows that the WAI result of 0.48 means that their general awareness of the impact of flooding is medium.

However, the WAI of the fishing group in Huong Phong is 0.40 and this means they consider flooding affecting livelihood activities of this group to be weak. They argue that flooding can also bring fish and shrimp resources in the natural environment for mobile fishing gear, and seasonal flooding is a favorable time for their exploitation of fisheries. In contrast, fishing groups in Quang An said that floods impact strongly on their livelihoods, and this is reflected in the WAI of this group (0.61). As rationale for this phenomenon, this fishermen group said that their fishing gear is fixed, such as corral net, bottom net, and so on, so a big flood will cause destruction of fixed fishing gear. So floods have a strong influence on their livelihoods.

The table also shows there is no significant difference between the level of awareness of flood impact on aquaculture households, as this group's WAI is 0.43–0.46 in Huong Phong and

TABLE 8.4   Fishery Community's Perception of Flood Affect on Livelihood

Level of Climate Change Effect	Quang An				Huong Phong				Total	
	AH		FH		AH		FH			
	N=29	%	N=31	%	N=27	%	N=18	%	N=105	%
Very strong	5	17.2	6	19.4	3	11.2	1	5.6	15	14.3
Strong	4	13.8	8	25.8	6	22.2	4	22.2	22	21
Medium	4	13.8	12	38.7	7	25.9	4	22.2	27	25.7
Weak	10	34.5	4	12.9	6	22.2	5	27.8	25	23.8
Very weak	6	20.7	1	3.2	5	18.5	4	22.2	16	15.2
WAI	0.43		0.61		0.46		0.40		0.48	
OA	M		S		M		W		M	

Note: Very weak (VW): 0.01–0.2; weak (W): 0.21–0.4; medium (M): 0.41–0.6; strong (S): 0.61–0.8; very strong (VS): 0.81–1.
OA, overall assessment; AH, aquaculture household; FH, fishing household.
Field survey, 2011.

Quang An, respectively, because most of the aquaculture ponds were in reinforced embankments that are higher than flood levels and flooding cannot affect their aquaculture operations.

8.3.2.3 Fishery Community's Perception of Precipitation Increase on Livelihood

Table 8.5 shows that the results of the WAI to assess the level of perception of local people about the impact of annual precipitation change for their livelihood in the Tam Giang lagoon water. The average WAI is 0.76, meaning that local people assess changing rainfall strongly affects their aquatic production activities.

TABLE 8.5   Fishery Community's Perception of Precipitation Increase on Livelihood

Level of Climate Change Effect	Quang An				Huong Phong				Total	
	AH		FH		AH		FH			
	N=29	%	N=31	%	N=27	%	N=18	%	N=105	%
Very strong	16	55.2	10	32.3	13	48.2	4	22.2	43	41
Strong	7	24.1	11	35.5	12	44.4	6	33.3	36	34.3
Medium	4	13.8	4	12.9	2	7.4	5	27.8	15	14.2
Weak	1	3.5	4	12.9	0	0	3	16.7	8	7.6
Very weak	1	3.5	2	6.4	0	0	0	0	3	2.9
WAI	0.81		0.69		0.85		0.65		0.76	
OA	VS		S		VS		S		S	

Note: Very weak (VW): 0.01–0.2; weak (W): 0.21–0.4; medium (M): 0.41–0.6; strong (S): 0.61–0.8; very strong (VS): 0.81–1.
OA, overall assessment; AH, aquaculture household; FH, fishing household.
Field survey, 2011.

Table 8.5 also shows that differences between fishing and aquaculture groups in their perceptions of the impact of rainfall on the livelihood of each group, but their WAI are quite the same at 0.81 in Quang An and 0.85 in Huong Phong, which means there is very strong influence.

Analysis of survey results and informant group discussions shows that precipitation reduces the salinity in aquaculture ponds. These conditions are not favorable for the aquatic species in the aquacultural pond; they just adapt to a brackish water environment. In these days of heavy rain, salinity in the ponds can be reduced to 00/00, causing massive death of fish and shrimp in ponds.

8.3.2.4 Fishery Community's Perception of Temperature Fluctuation on Livelihood

According to the survey results shown in Table 8.6, which represents average results of WAI as 0.74, most local fishermen were aware of the importance of weather temperatures for their livelihood activities, and this means that temperature strongly affects the production of aquatic activities. However, there is a difference between fishing and aquaculture groups in Quang An and Huong Phong.

In Huong Phong, the WAI of the aquaculture group is 0.83, which is very strong influence, while the WAI of the fishing group is only 0.57, which means average effects. The main reason for this situation is the change in temperature in the pond (both cold and hot, sudden or prolonged). Extremely hot temperature will cause water stratification in the pond, which leads to aquatic species not becoming adapted. The temperature is too cold in the rainy season, and this can lead to the phenomenon of mass death of fish and shrimp in aquaculture ponds.

However, the fishing households think only the extreme cold temperature affects their livelihoods. If it is cold they cannot use mobile fishing gear outside of the lagoon water, especially at night when the temperature drops too low. This is shown in Table 8.5, where the WAI of the fishing group is 0.57 in Huong Phong and 0.69 in Quang An.

TABLE 8.6   Fishery Community's Perception of Temperature Fluctuation on Livelihood

Level of Climate Change Effect	Quang An				Huong Phong				Total	
	AH		FH		AH		FH			
	N=29	%	N=31	%	N=27	%	N=18	%	N=105	%
Very strong	13	44.8	7	22.6	14	51.9	2	11.1	36	34.3
Strong	10	34.5	12	38.7	9	33.3	4	22.2	35	33.3
Medium	5	17.2	10	32.3	3	11.1	9	50	27	25.7
Weak	1	3.5	2	6.4	1	3.7	3	16.7	7	6.7
Very weak	0	0	0	0	0	0	0	0	0	0
WAI	0.80		0.69		0.83		0.57		0.74	
OA	S		S		VS		M		S	

Note: Very weak (VW): 0.01–0.2; weak (W): 0.21–0.4; medium (M): 0.41–0.6; strong (S): 0.61–0.8; very strong (VS): 0.81–1.
OA, overall assessment; AH, aquaculture household; FH, fishing household.
Field survey, 2011.

8.4 CONCLUSIONS AND RECOMMENDATIONS

8.4.1 Conclusions

In Tam Giang lagoon, the weather and environment have been changing due to climate change. The manifestations of climate change in this lagoon are the increasing number and intensity of storms, temperature and precipitation fluctuation, flood increase, and sea level rise.

Fishermen in Tam Giang lagoon are aware of the impact of climate change on their lives, livelihoods, and the resources of the lagoon. People realize that climate change has been leading to aquatic resources degradation in study areas, including both Quang An and Huong Phong communes, threatening the fishermen's livelihood and communities who directly depend on lagoon natural resources.

According to people's perception, the manifestation of climate change impacts is different for each livelihood activity of the fishery household. In general, climate change seriously impacts the livelihoods of fishermen; however in some cases, climate change benefits the fishermen group, such as floods that bring benefits for the mobile fishing group.

8.4.2 Recommendations

To improve and promote the role of fishermen to cope with climate change in Tam Giang lagoon, through the research results, the following recommendations are made.

1. Local government and fishery communities have to receive needed assistance and training programs to improve their knowledge and efficiency to adapt to climate change affects. Government and scientists need to promote the search for alternative livelihood solutions for fishermen, especially for the mobile fishing group who is vulnerable to the effects of climate change.
2. Researchers should focus on the perception of fishermen when they conduct research related to climate change in Tam Giang lagoon. In addition, the application of indigenous knowledge of the people in building effective solutions to climate change adaptation should be made.
3. Improved information and communication networks between government, scientists, and fishery communities to provide early warning to fishermen can deal effectively with the complicated weather fluctuation and share information and experience on adaptation to the effects of climate change.

References

Center Hydrometeorology Prediction, 2009. Thua Thien Hue Hydrometeorology Prediction. CHP annual report, Statistic Publisher, Hue, Vietnam.

Cu, N.H., 1996. Geological features of Tam Giang—Cau Hai lagoon. In: National Conference Report on the Lagoon. Hue, Vietnam.

Huong Phong, 2011. Aquaculture Report of Huong Phong Commune in 2010. Huong Phong annual report, Commune People's Committee, Huong Phong, Thua Thien Hue.

MONRE, 2009. The Scenario of Climate Change and Sea Level Rise in Vietnam. Ministry of Natural Resource and Environment, from http://www.mcdvietnam.org/Uploaded/admins/360%20do/Climate%20change/Tai%20lieu/Kich%20ban%20Bien%20doi%20khi%20hau.pdf.

NAV, 2010. Situation Analysis on Climate Change in Thua Thien Hue Province, Vietnam. Nordic Assistant to Vietnam annual report, Nordic Assistance to Vietnam Project, Hue, Vietnam.

Nguyen, B.Q.V., 2008. Research in modeling project on community based management for small-scale fisheries in Thua Thien Hue, Hue University of Sciences. Hue univ. 30, 124–145.

Nguyen, T., Nguyen, D.N., 2010. Effect of climate change on Perfurm River basin, Thua Thien Hue Province. Hue Univ. Sci. 58, 231–350.

Nguyen Viet, 2008. Natural disasters in Thua Thien Hue province and the official general avoidance measures. Annual report, Hydrometeorology Prediction of Thua Thien Hue. Center for Hydrometeorology Prediction of Thua Thien Hue province.

Quang An, 2011. Aquaculture Report of Quang An Commune in 2010. Quang An annual report, Commune People's Committee, Huong Phong, Thua Thien Hue.

Thua Thien Hue CPC, 2010. Scenarios of Climate Change in Thua Thien Hue Province. Thua Thien Hue Statistic Publisher, Hanoi, Vietnam.

Thua Thien Hue DARD, 2010. Tam Giang—Cau Hai Lagoon Hydrometeorology Prediction. Yearly report, the Thua Thien Hue Department of Agriculture and Rural Development Publisher.

Thung, D.C., 2006. Natural Resources and Environment in Tam Giang—Cau Hai Lagoon. Synthesis report, IMOLA, Hue, Vietnam.

Tuyen, T.V., Armitage, D., Marschke, M., 2010. Livelihoods and co-management in the Tam Giang lagoon, Vietnam. Ocean Coast. Manag. 53, 327–335.

Reconciling Science and Indigenous Knowledge in Selecting Indicator Species for Forest Monitoring

T.D. Ngo*, E.L. Webb†

*Hue University of Agriculture and Forestry, Hue, Vietnam †National University of Singapore, Singapore, Singapore

9.1 INTRODUCTION

Ecologists and conservation practitioners have increasingly utilized indicator species for environmental monitoring and assessment. A variety of taxonomic groups are being used to assess and monitor diverse parameters such as biodiversity (Andersen et al., 2004; Cardoso et al., 2004), forest condition (Canterbury et al., 2000), logging impacts (Cleary, 2004), and ecological and socioeconomic changes of integrated conservation and development programs (Kremen et al., 1998). Research continues to improve methods of indicator species development, based on data collected through in-depth field research (Dufrêne and Legendre, 1997; McGeoch, 1998; Noss, 1999; Grove, 2002; Kati, 2003; Andersen et al., 2004; Schmidt et al., 2006).

Plant species may be good indicators for forest disturbance and forest restoration (McLachlan and Bazely, 2001), for the functioning of ecosystems (Cole, 2002), and impacts of human activities on forest communities (Dale et al., 2002). Plant species can be used as indicators for disturbance because they are easily sampled and stored (Hammond, 1994), taxonomically well-known and abundant (Stork, 1994), sufficiently sensitive to provide an early warning (Noss, 1990), distributed over a range of habitats or environments (Faith and Walker, 1996), and have economic importance for local people (Kremen et al., 1998).

Development of indicator species usually employs intensive survey methods that require well-developed researcher qualifications. Moreover, most indicators have been developed by scientists outside the area being monitored and have often not integrated local, indigenous expertise, otherwise known as local ecological knowledge (Salam et al., 2006; Hambly, 1996), into the indicator set. However, there may be substantial benefits from including local knowledge in the development of ecological indicators. Local people may have extensive observational experience that can complement quantitatively derived methods, thereby saving both time and money. They can also provide historical contexts of the area being monitored, to better understand the dynamics of the ecosystem to be monitored. Local ecological knowledge has enormous potential to contribute to collaborative conservation and monitoring (Steinmetz et al., 2006; Moller et al., 2004).

In Vietnam, local people are increasingly being recognized as "forest owners" through the process of forest allocation (Sikor and Apel, 1998), where the rights to access and harvest forest materials on specifically delineated parcels of degraded land and natural forest are being given to recipient households for 50-year periods (Dung and Webb, in press). Along with the rights to use the forest, recipients are obligated to manage and monitor the forest condition and to maintain ecological integrity. Thus, local people in Vietnam need to be involved and proactive in forest management. Locally relevant ecological indicators derived through a combination of traditional survey techniques and local interviews could therefore contribute to long-term sustainable conservation and management of forest in Vietnam.

In the district of Nam Dong in the central province of Thua Thien Hue, local people have a long history of forest access (Thang, 2004) and therefore could contribute to the development of forest monitoring indicators. Swidden agriculture and nontimber forest product collection have provided local people with extensive experience on plant species that help them decide where and when to swidden, and which crops are suitable for particular sites. Moreover, through forest accessing local people have gained a high familiarity with plant species variation across forest disturbance levels. Some herbaceous plants can be used to identify wildlife habitats for hunting (Hong, 2002). With high forest cover and a highly dynamic landscape in Nam Dong (eg, logging, road development, and swidden agriculture; Thiha et al., 2007), plant indicators may have potential to be used for rapid assessment of forest condition and long-term monitoring in forests of Nam Dong.

This study developed a list of indicator species for monitoring forest disturbance by using both local knowledge and quantitative data. These two sources of data were expected to complement each other to determine which plant species are characteristic (ie, indicators) of forests under different levels of disturbance. Using both data sources of indicator species were expected to be both statistically sound and easily applied by local people in monitoring the disturbance level of their allocated forest, and by state agencies in monitoring unallocated forest.

9.2 MATERIALS AND METHODS

9.2.1 Study Area

Nam Dong District is located in the southwestern part of Thua Thien Hue Province in Central Vietnam (Fig. 9.1). This district lies to the north of low mountains extending from the Annamite Mountains on the Laos border to the East Sea. The district ranges from 300 to 1700 m above sea level (asl), and approximately 65% of the district is covered with seasonal evergreen broadleaf forest (Rundel, 1999; Tordoff et al., 2003). Most of the forests below 400 m asl have been converted into plantations of *Hevea brasiliensis* for rubber latex or *Acacia mangium* for pulp. Followed Decree 02/CP of the government on forest land allocation in

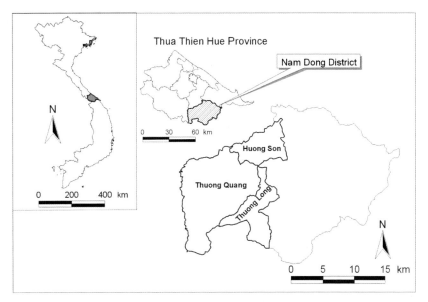

FIG. 9.1 Map of Central Vietnam showing Nam Dong District and the communes of this study.

1994, Nam Dong authorities had allocated 9458 ha of forest land to households and groups of households of which 6700 ha was natural forest by the year 2005 (Banh, 2006).

Three communes of Nam Dong District were selected for this study: Thuong Quang, Thuong Long, and Huong Son (Fig. 9.1). Most of people in these communes are of Katu ethnicity and were resettled from the mountainous border with Laos after reunification in 1975, along with a group of Kinh majority people who were resettled from central coastal zones to Nam Dong. Major livelihood activities were swidden agriculture and forest product harvesting until 1990 when government policy placed restrictions on slash-and-burn practices in only certain specified areas. However, forest products are routinely harvested for local and regional markets, particularly rattan, bamboo, animals, leaves of *Rhapis laosensis* (for making conical hats), and flowers of *Thysanolaena maxima* (for making brooms) (Thang, 2004). Illegal timber harvesting also occurs on a small scale (Thang, 2004; Thiha et al., 2007).

Since the year 2000, the government has been allocating natural forest for long-term management. Allocated forests are characterized by variable levels of disturbance and recovery. We classified forest types for this study into three groups, based on a combination of the government class system and local description (Table 9.1). Relatively intact forest (RIF) was the forest that had a closed-canopy, well-defined stratification, and a majority of large trees with diameter greater 20 cm. RIF had been logged in the past but had recovered. Selectively logged forest (SLF) was the forest that had been heavily logged and therefore had significantly altered structures. Shrubs, vines, and bamboo accounted for more than 50% of forest cover in SLF. Swidden forest (SWF) was the forest that regenerated after swidden agriculture or clear-cutting harvest. SWF was dominated by light-demanding, fast-growing tree species (pioneers). The majority of allocated forests have been RIF and SLF, with small areas of SWF allocated in Thuong Quang for plantation or upland crop cultivation (cassava, maize, and upland rice).

TABLE 9.1   Classification of Forest by Disturbance Factors

Local System	Government System	Combined System
Rung non: Regenerated forest after swidden agriculture	*Category: IIA, IIB* – Forest after swidden (IIA) or after clear cutting (IIB) – Average DBH below 20 cm	*Swidden forest (SWF)* – Regenerated from swidden agriculture or clear-cutting harvest – Dominated by light-demanding, fast-growing tree species (pioneer species)
Rung bim: Forest after selected logging with small trees, shrubs, vines, and bamboo	*Category: IIIA1* – Heavily logged secondary forest – Insufficient seedlings (IIIA1.1) or sufficient seedlings (IIIA1.2)	*Selectively logged forest (SLF)* – Completely destroyed or transformed structure with big gaps – Few large trees remaining – Shrubs, vines, and bamboo occupied >50% of forest cover
Rung gia: Undisturbed forest or very less disturbed with large timber tree species	*Category: IIIA2, IIIA3* – Secondary forest disturbed by logging but substantially recovered – Average DBH in range of 20–30 cm (IIIA2) or >35 cm (IIIA3)	*Relatively intact forest (RIF)* – Closed-canopy forest – Majority of large trees with DBH >20 cm – Multistrata forest

The government system on disturbed forest types followed "Procedures for Forest Business Designing" (QPN6-84) issued by the Minister of Forestry (Ministry of Forestry, MOF, 1984).

9.2.2 Methodology

We collected two types of data to construct a list of forest disturbance indicator species. The first one was qualitative data obtained from local knowledge. The perceptions of local people on indicators for three disturbed forest types (RIF, SLF, and SWF) were collected through group meetings and individual interviews. A group meeting was held in each commune with villagers who were experienced with forest uses. The local system of categorizing forest disturbance types was combined with the government classification system (Table 9.1). Local people were then asked to list all plant species that occurred exclusively in a particular disturbed forest type. We focused on life forms larger than seedlings because people tended not to associate seedlings of most species with particular habitat types. This reference plant species list was then used to design a questionnaire for individual interviews.

Village heads and commune staff were consulted to help purposively select 118 interviewees who had a long residence in the village, had practiced swidden agriculture, or had long-term experience in uses of forest resources, and who had knowledge on species that could be used as forest disturbance indicators. After all interviews, species were then ranked for each forest type based on the percentage of respondents who selected the species for that forest (ResVal). If the selected species was an indicator for more than one forest type, it was listed in only the forest type where it had the highest ResVal.

A second data type for constructing a list of indicator species was obtained through quantitative forest plots. We established 110 temporary plots in the three forest types in three communes (Table 9.2). All plots were randomly located within allocated forest areas,

TABLE 9.2   Brief Information on Surveyed Plots in Three Disturbed Forest Types of Three Communes in Nam Dong District, Central Vietnam

| | Number of Forest Plots | | | | | | |
Forest Types	Thuong Quang	Thuong Long	Huong Son	Total	Area (ha)	No. of Families	No. of Species
Relatively intact forest (RIF)	21	30	0	51	1.60	65	156
Selectively logged forest (SLF)	11	0	30	41	1.29	53	136
Swidden forest (SWF)	18	0	0	18	0.57	34	57
Total	50	30	30	110	3.46	72	184

so that results would be immediately demonstrable to local people. Each circular plot had a radius of 10 m with a 3 m radius plot nested within (Ostrom et al., 2004). In the 10 m plot, we recorded the species name, diameter, and height of each tree with a diameter at breast height (DBH) greater than or equal to 10 cm. In the 3 m plot, we recorded the same information for saplings with a DBH in the range of 2.5–10 cm. Total survey area for trees was 3.46 ha, and for saplings total area was 3110 m^2 (Table 9.2). We did not survey woody seedlings, shrubs, or woody climbers with a DBH less than 2.5 cm or a height less than 1.0 m because they were not comparable to the locally derived indicator list, which focused on life forms larger than seedlings. The plot surveys at each commune were conducted with at least two local people from the group interviews. Taxonomy followed *An Illustrated Flora of Vietnam* (Ho, 2000) and the *Plant Name Index of Vietnam* (Institute of Ecology and Biological Resources, IEBR and Center for Resources and Environmental Studies, CRES, 2003).

From plot data, the indicator value (IndVal) of each species was calculated using the method of Dufrêne and Legendre (1997). This method combined information on the concentration of species abundance in a particular forest type (specificity) and the faithfulness of occurrence of a species in a particular forest type (fidelity).

$$\text{Specificity measure: } A_{ij} = \text{Nindividuals}_{ij} / \text{Nindividuals}_i$$

where Nindividuals$_{ij}$ is the mean number of species i across sites of forest type j, and Nindividuals$_i$ is the sum of the mean numbers of individuals of species i over three forest types (RIF, SLF, and SWF).

$$\text{Fidelity measure: } B_{ij} = \text{Nplots}_{ij} / \text{Nplots}_j$$

where Nplots$_{ij}$ is the number of plots in forest type j where species i is present, and Nplots$_j$ is the total number of plots in that forest type.

$$\text{Indicator value: IndVal}_{ij} = A_{ij} \times B_{ij} \times 100$$

where IndVal$_{ij}$ is the indicator value of species i in forest type j. Thus, each species was given an IndVal for each forest type. The final listed IndVal of species i was the highest value among these three IndVals.

The IndVal may range from 0 (no indication) to 100 (perfect indication). Perfect indication means that the presence of a species points to a particular forest type without error, at least with the available data set. To test whether the observed IndVal of a species in a certain forest disturbance type was significantly higher than would be expected based on a random distribution, the observed IndVal was compared with 1000 randomly generated IndVals. If the observed IndVal of a species in a forest disturbance type fell within the top 5% of the random IndVals, it was considered to be a significant indicator for that forest type (McCune and Grace, 2002). PC-ORD (version 4.41) was used to calculate both IndVal and p-value of each species (McCune and Mefford, 1999).

We compared the results of interviews and plots to construct a final list of indicators. The final list of indicator species was selected based on two major criteria: overlapping (ie, local knowledge and plot data showed the same indicator species for a particular forest type) and ranking (ie, rank of a particular indicator species based on its ResVal or IndVal). Two groups of indicator species were selected. The "probable indicator species" included all species that were significant indicators for both local knowledge and plot data. The "possible indicator species" were those that appeared in only one list, but that had a rank in the top three for that list.

9.3 RESULTS

9.3.1 Indicator Species From Interviews With Local People

Thirty-two species were identified by local people as indicators for three disturbed forest types, with 15 for RIF, 9 for SLF, and 8 for SWF (Table 9.3). Top indicator species for RIF were *Hopea pierrei*, *Erythrophleum fordii*, and *Parashorea stellata*. Three top indicator species for SWF were *Thysanolaena maxima*, *Mallotus paniculatus*, and *Imperata cylindrica*. Top indicators for SLF were *Horsfieldia amygdalina*, *Ampelocalamus patellaris*, *Gardenia annamensis*, and *Wrightia annamensis*. The top SLF indicators had ResVals that were lower than RIF or SWF.

Among 32 indicators, tree species accounted for 68.8%; palm, grass, and herbs shared 6.3% for each group; shrub, bamboo, woody climbers, and vine had 3.1% for each life form. Most indicators for RIF and SLF were trees. Herbs and grasses made up a majority of SWF indicator species.

9.3.2 Indicator Species From Forest Plot Data

Analysis of plot data yielded 21 indicator species for RIF, 30 for SLF, and 15 for SWF (Table 9.4). Among the top three indicator species for each forest type, indicators for SWF had the highest IndVal followed by SLF and RIF. In SWF, top indicator species were *Mallotus paniculatus* and *Macaranga denticulata*. Indicator species for SLF were *Scaphium lychnophorum*, *K. pierrei*, and *Croton cascarilloides*. Indicator species for RIF were *Gironniera subaequalis*, *Palaquium annamense*, and *Syzygium* sp.

Trees were the dominant life form among plot-based indicator species. Large trees (>30 m tall) indicating RIF were *Palaquium annamense*, *Madhuca pasquieri*, *Hopea pierrei*, and *Canarium bengalense*. One palm species (*Rhapis laosensis*) was included in RIF indicators. Fifty percent

TABLE 9.3 Indicator Species for Three Disturbed Forest Types Based on Analysis of Local Interview Data

Rank	Relatively Intact Forest (RIF)			Selectively Logged Forest (SLF)			Swidden Forest (SWF)		
	Species	ResVal	Life Form	Species	ResVal	Life Form	Species	ResVal	Life Form
1	*Hopea pierrei*	69.5	L-tree	*Horsfieldia amygdalina*	18.6	L-tree	*Thysanolaena maxima*	67.8	Grass
2	*Erythrophloeum fordii*	45.8	L-tree	*Ampelocalamus patellaris*	16.1	Bamboo	*Mallotus paniculatus*	59.3	M-tree
3	*Parashorea stellate*	42.4	L-tree	*Gardenia annamensis*	15.3	S-tree	*Imperata cylindrica*	52.5	Grass
4	*Sindora tonkinensis*	34.7	L-tree	*Wrightia annamensis*	15.3	S-tree	*Melastoma candidum*	38.1	Shrub
5	*Calamus* spp.	30.5	Palm	*Gnetum latifolium*	11.9	W-climber	*Ipomea* sp.	19.5	Vine
6	*Rhapis laosensis*	28.8	Palm	*Garcinia cochinchinensis*	10.2	M-tree	*Crassocephalum crepidioides*	17.8	Herb
7	*Gironniera subaequalis*	22.9	M-tree	*Trema orientalis*	7.6	S-tree	*Macaranga denticulata*	16.9	S-tree
8	*Syzygium* sp.	21.2	S-tree	*Gonocaryum maclurei*	4.2	M-tree	*Eupatorium odoratum*	5.9	Herb
Total	15			9			8		

L-tree: large tree (>30 m high); *M-tree*: medium tree (15–30 m high); *S-tree*: small tree (<15 m high); *W-climber*: woody climber.
Life form of plant species followed "The Names of Forest Plants in Vietnam" (Ministry of Agriculture and Rural Development, MARD, 2000).

TABLE 9.4 Indicator Species From Plot Survey in Three Disturbed Forest Types

Rank	Relatively Intact Forest (RIF)				Selectively Logged Forest (SLF)				Swidden Forest (SWF)			
	Species	IV	p-Value	Life Form	Species	IV	p-Value	Life Form	Species	IV	p-Value	Life Form
1	Gironniera subaequalis	49.4	0.001	M-tree	Scaphium lychnophorum	61.2	0.001	M-tree	Mallotus paniculatus	77.8	0.001	M-tree
2	Palaquium annamense	43.9	0.001	L-tree	Knema pierrei	54.4	0.001	L-tree	Macaranga denticulata	70.2	0.001	S-tree
3	Syzygium sp.	39.4	0.009	S-tree	Croton cascarilloides	50.6	0.001	Shrub	Scleria levis	55.6	0.001	Grass
4	Canarium bengalense	38.1	0.001	L-tree	Castanopsis sp.	44.4	0.001	L-tree	Melastoma candidum	28.4	0.012	Shrub
5	Schefflera octophylla	34.2	0.003	M-tree	Barringtonia macrostachya	44.2	0.001	M-tree	Macaranga sp.	25.1	0.001	S-tree
6	Acronychia pedunculata	30.9	0.005	S-tree	Pometia pinnata	41.7	0.001	L-tree	Thysanolaena maxima	22.2	0.001	Grass
7	Madhuca pasquieri	28.0	0.004	L-tree	Horsfieldia amygdalina	38.5	0.002	L-tree	Trema orientalis	22.2	0.001	S-tree
8	Cinnamomum parthenoxylum.	27.2	0.035	L-tree	Gardenia annamensis	35.9	0.002	S-tree	Phrynium capitatum	22.2	0.002	Grass
9	Cratoxylum pruniflorum	25.9	0.001	S-tree	Pometia sp.	33.8	0.002	L-tree	Anthocephalus indicus	16.7	0.007	L-tree
10	Hopea pierrei	25.0	0.011	L-tree	K. poilanei	32.8	0.004	S-tree	Ficus sp.	16.7	0.007	L-tree
11	Cryptocaria lenticellata	24.2	0.007	M-tree	Endospermum chinense	32.3	0.001	L-tree	Saurauia roxburghii	15.3	0.013	S-tree
12	Elaeocarpus griffithii	22.9	0.008	L-tree	Gonocaryum maclurei	31.3	0.015	M-tree	Rubus alcaefolius	11.1	0.028	Shrub
13	Bouea oppositifolia	22.2	0.024	M-tree	Nephelium sp.	30.5	0.024	L-tree	Eupatorium odoratum	11.1	0.033	Herb
14	Quercus sp.	20.8	0.018	L-tree	Artocarpus rigidus	29.8	0.008	L-tree	Averrhoa carrambola	11.1	0.036	M-tree
15	Gordonia balansae	20.2	0.015	S-tree	Parrashorea stellata	27.6	0.01	L-tree	Xylopia vielana	9.4	0.039	M-tree
Total	21				30				15			

L-tree: large tree (>30m high); M-tree: medium tree (15–30 m high); S-tree: small tree (<15m high); W-climber: woody climber. Only nonseedling life forms with p-value <0.05 are listed. Life form of plant species followed "The Names of Forest Plants in Vietnam" (Ministry of Agriculture and Rural Development, MARD, 2000).

of indicator species for SLF were large trees (eg, *K. pierrei*); 20% were medium trees (25–30 m tall) (eg. *Scaphium lychnophorum*); 10% were shrubs (eg, *Croton cascarilloides*); and one woody climber (*Strophanthus divaricatus*). Indicator species for SWF included all life forms such as trees (*Mallotus paniculatus, Macaranga denticulata*), grass (*Scleria levis, Thysanolaena maxima*), shrub (*Melastoma candidum*), and herb (*Eupatorium odoratum*).

9.3.3 Final List of Indicator Species

Comparing the interview data with the plot data returned 13 probable indicators (Table 9.5) and 7 possible indicators (Table 9.6). *Hopea pierrei, Gironniera subaequalis*, and *Palaquium annamense* appeared to be the strongest probable indicators for RIF because they had both high ResVal and high IndVal. Indicator species for SLF included *Horsfieldia*

TABLE 9.5 Probable Indicator Species for Relatively Intact Forest (RIF), Selectively Logged Forest (SLF), and Swidden Forest (SWF) in Nam Dong District of Central Vietnam

Forest Type	Species Name	Family	Life Form	Local Knowledge		Plot Data	
				ResVal	Rank	IndVal	Rank
RIF	*Hopea pierrei*	DIPTEROCARPACEAE	L-tree	69.5	1	25.0	10
	Gironniera subaequalis	ULMACEAE	M-tree	22.9	7	49.4	1
	Palaquium annamense	SAPOTACEAE	L-tree	17.8	9	43.9	2
	Syzygium sp.	MYRTACEAE	S-tree	21.2	8	39.4	3
	Rhapis laosensis	ARECACEAE	Palm	28.8	6	11.7	21
	Madhuca pasquieri	SAPOTACEAE	L-tree	13.6	10	28.0	7
SLF	*Horsfieldia amygdalina*	MYRISTICACEAE	L-tree	18.6	1	38.5	7
	Gardenia annamensis	RUBIACEAE	S-tree	15.3	3	35.9	8
	Garcinia cochinchinensis	CLUSIACEAE	M-tree	10.2	6	26.2	16
SWF	*Mallotus paniculatus*	EUPHORBIACEAE	M-tree	59.3	2	77.8	1
	Macaranga denticulate	EUPHORBIACEAE	S-tree	16.9	7	70.2	2
	Thysanolaena maxima	POACEAE	Grass	67.8	1	22.2	6
	Melastoma candidum	MELASTOMATACEAE	Shrub	38.1	4	28.4	4

Respondent values (ResVal) and indicator values (IndVal) were listed from local interview data and plot survey data, respectively.

TABLE 9.6   Possible Indicator Species for Relatively Intact Forest (RIF), Selectively Logged Forest (SLF), and Swidden Forest (SWF) in Nam Dong District of Central Vietnam

Forest Type	Species Name	Family	Life Form	Local Knowledge		Plot Data	
				ResVal	Rank	IndVal	Rank
RIF	*Erythrophleum fordii*	CAESALPINIACEAE	L-tree	45.8	2	–	–
SLF	*Scaphium lychnophorum*	STERCULIACEAE	M-tree	–	–	61.2	1
	Knema pierrei	MYRISTICACEAE	L-tree	–	–	54.4	–
	Croton cascarilloides	EUPHORBIACEAE	Shrub	–	–	50.6	3
	Ampelocalamus patellaris	POACEAE	Bamboo	16.1	2	–	–
SWF	*Scleria levis*	POACEAE	Grass	–	–	66.1	3
	Imperata cylindrica	POACEAE	Grass	52.5	3	–	–

Respondent values (ResVal) and indicator values (IndVal) were listed from local interview data and plot survey data, respectively.

amygdalina, Gardenia annamensis, and *Garcinia cochinchinensis*. However, both ResVals and IndVals of these indicators were not high. SWF indicators were *Mallotus paniculatus, Macaranga denticulata, Thysanolaena maxima*, and *Melastoma candidum*. In particular, *Mallotus paniculatus* appeared to be the strongest indicator with high ResVal and IndVal.

Trees accounted for the largest portion in all probable indicators (10 species). One SWF indicator (*Thysanolaena maxima*) was a grass. One palm species was an indicator for RIF (*Rhapis laosensis*).

Possible indicator species included seven species for the three forest types. RIF indicator was *Erythrophleum fordii*. Four indicators for SLF were *Scaphium lychnophorum, K. pierrei, Croton cascarilloides*, and *Ampelocalamus patellaris*. Indicators for SWF were *Imperata cylindrica* and *Scleria levis*. Possible indicators included four tree species, two grass species, one bamboo species, and one shrub species.

9.4 DISCUSSION

The fact that local people focused on useful species is a limitation of the method of this study. Local people's responses on indicator species were very much dependent on particular uses of those species (Kremen et al., 1998; Matavele and Habib, 2000) and frequency of access to a forest type. As good timber tree species were no longer available for local uses in the SLF due to selected logging, few responses on potential indicator species for this forest were recorded. In addition, local people got familiar with species composition in the SWF because these areas were transformed to agricultural crops (eg, cassava, maize) annually or forest plantation (*Acacia mangium*). This could be the reason that ResVals of SLF indicator species were much lower than those of RIF and SWF in local interviews.

Several other options are available for selection of indicators to monitor different levels of forest disturbance. For example, using higher taxa (genera or families) would be more suitable for identifying indicators of highly diverse forest due to a highly closed link between species richness and generic richness (Balmford et al., 1996). In our study, the top three IndVals were lower in RIF than in SWF and SLF partly because a diverse forest tends to have lower species dominance than a species-poor forest (Whittaker, 1965; Brooks and Matchett, 2003; Dangles and Malmqvist, 2004). In addition, Balmford et al. (1996) also confirmed that field work using woody plant genera and families instead of species reduced survey costs by a minimum of 60–85%, respectively. Besides, using dead logs as potential indicators (Grove, 2002) could be another alternative for SLFs.

The indicator species in this study were recorded in other research in similar forest conditions. Rundel (1999) found that species of *Madhuca pasquieri*, *Hopea pierrei*, and *Gironniera subaequalis* were common in wet evergreen forest of central northern Vietnam. Slik et al. (2003) showed that species of two genera (*Mallotus* and *Macaranga*) were indicators for highly disturbed forest in East Kalimantan, Indonesia. Local people in a Himalayan region often practise "honeyem," a process of using seedlings of *Macaranga denticulata* and *Mallotus tetracoccus* to increase soil fertility, check soil erosion, and produce fuel wood and timber around rice fields (Aumeeruddy-Thomas and Shengji, 2003). *Scaphium lychnophorum* often occurs in gaps of moderately disturbed forest (Chan and Huyen, 2000; Hy, 2005). Thus, our list of probable indicator species agrees well with previous research.

The occurrence of a particular species in a specific stage of forest succession may be the reason for the absence of several interview-derived indicators in the list of plot-derived indicators. *Crassocephalum crepidioides* and *Eupatorium odoratum* appeared only at the first stage of succession in the fallow period of swidden agriculture practice (Schmidt-Vogt, 1999; Swamy et al., 2000). Similarly, *Imperata cylindrica* occurs in Asia and Africa after shifting agriculture, and disappears with postdisturbance canopy closure (Howard, 2005). Therefore, these species were rarely seen in canopy-closed forest plots in Nam Dong. Similarly, the overexploitation of two species, *Erythrophleum fordii* and *Parashorea stellata*, due to the high demand of commercial timber (Tordoff et al., 2003; Lan et al., 2002; Dung, 1996) caused the absence of adult trees in surveyed forest. As a result, these two species were not found among the top three indicator species of RIF plot data.

Indicators found in this study can be applied in stress-oriented or predictive monitoring (NRC, 1995a). In this type of monitoring, there are two contexts in which indicator species can be used. First, early warning of forest degradation can be made based on the increasing incidence of SWF indicators and the decrease of RIF indicators in a forest. Second, the recovery of degraded forest can be predicted through the increased RIF or decreased SWF indicator species in a currently surveyed forest. In this type of monitoring, a cause and effect relationship is known between indicator species and their occurrence in a specific forest disturbance.

Seedlings were not included in plot data to make the plot data set comparable with local knowledge. Seedlings, however, could be "indirect" indicators for forest management practices. For example, seedlings of indicator species for RIF (*Hopea pierrei*, *Madhuca pasquieri*, and *Palaquium annamense*) could be used as predictors of succession or forest health if favorable conditions are given for these seedlings' growth. Similarly, the presence of seedlings of these species in selected logging forest can predict the future direction of the current forest type to the RIF through gap-enhanced recruitment (Dupuy and Chazdon, 2006; Steven, 1994). Enhancement of local awareness about seedlings of these species through forest rehabilitation can lead to the use of seedlings as indicators for level of forest disturbance. Future research should identify which tree species with easily recognizable seedlings and saplings could serve as indicator species in Nam Dong.

Sheil (2001) suggested that biodiversity monitoring activities could hinder conservation in tropical developing countries if they divert scarce resources away from fundamental management priorities. This "dilemma," however, can be resolved through a more cost-effective monitoring protocol by using indicator species. In this monitoring system, local knowledge and quantitative forest data are packaged together in three steps of selection and application indicator species. First, a list of indicator species derived from local interviews could help to screen potential indicators commensurate with local interests. Second, these species can then be cross-checked by quantitative plot or transect surveys. Finally, local people can integrate these indicator species in monitoring a section of the community forest. This combination helps to both reduce the cost of monitoring (which is normally done by a third party) and improve local awareness on conservation through their monitoring group work.

References

Andersen, A.N., Fisher, A., Hoffmann, B.D., Read, J.L., Richards, R., 2004. Use of terrestrial invertebrates for biodiversity monitoring in Australian rangelands, with particular reference to ants. Aust. Ecol. 29, 87–92.

Aumeeruddy-Thomas, Y., Shengji, P., 2003. Applied Ethnobotany: Case Studies From the Himalayan Region. People and plants working paper 12, WWF, Godalming, UK.

Balmford, A., Jayasuriya, A.H.M., Green, M.J.B., 1996. Using higher-taxon richness as a surrogate for species richness: II. Local applications. Proc. R. Soc. Lond. B 263, 1571–1575.

Banh, T. H., 2006. Review of Total Areas of Forest Land Allocated to Local Communities in Thua Thien Hue Province. Report to Green Corridor Project, Hue City.

Brooks, M.L., Matchett, J.R., 2003. Plant community patterns in unburned and burned blackbrush (*Coleogyne ramosissima* Torr.) shrublands in the Mojave Desert. West. N. Am. Naturalist 63, 283–298.

Canterbury, G.E., Martin, T.E., Petit, D.R., Petit, L.J., Bradford, D.F., 2000. Birds community and habitat as ecological indicator of forest condition in regional monitoring. Conserv. Biol. 14, 544–558.

Cardoso, P., Sinra, I., de Oliveira, N.G., Serrano, A.R.M., 2004. Higher taxa surrogates of spider (*Araneae*) diversity and their efficiency in conservation. Biol. Conserv. 117, 453–459.

Chan, L.M., Huyen, D.T., 2000. Thuc Vat Rung (Textbook of Forest Trees). Forestry University of Vietnam, Ha Tay (in Vietnamese).

Cleary, D.F.R., 2004. Assessing the use of butterflies as indicators of logging in Borneo at three taxonomic level. J. Econ. Entomol. 97, 429–435.

Cole, C.A., 2002. The assessment of herbaceous plant cover in wetlands as indicator of function. Ecol. Indic. 2, 287–293.

Dale, V.H., Beyeler, S.C., Jackson, B., 2002. Understory vegetation indicator of anthropogenic disturbance in longleaf pine forest at Fort Benning, Georgia, USA. Ecol. Indic. 1, 155–170.

Dangles, O., Malmqvist, B., 2004. Species richness-decomposition relationships depend on species dominance. Ecol. Lett. 7, 395–402.

Dufrêne, M., Legendre, P., 1997. Species assemblages and indicator species: the need for a flexible asymmetrical approach. Ecol. Monogr. 67 (3), 345–366.

Dung, V.V. (Ed.), 1996. Vietnam Forest Trees. Agricultural Publishing House, Hanoi, p. 788.

Dung, N.T., Webb, E.L., in press. Incentives of the forest land allocation process: implications for forest management in Nam Dong District, Central Vietnam. In: Webb, E.L., Shivakoti, G.P. (Eds.), Decentralization, Forests and Rural Communities: Policy Outcomes in South and Southeast Asia. Sage Press, Delhi.

Dupuy, J.M., Chazdon, R.L., 2006. Effects of vegetation cover on seedling and sapling dynamics in secondary tropical wet forests in Costa Rica. J. Trop. Ecol. 22, 65–76.

Faith, D.P., Walker, P.A., 1996. How do indicator groups provide information about the relative biodiversity of different sets of areas?: on hotspots, complementary and pattern-based approaches. Biodivers. Lett. 3, 18–25.

Grove, S.J., 2002. Tree basal area and dead wood as surrogate indicators of saproxylic insect faunal integrity: a case study from the Australian lowland tropics. Ecol. Indic. 1, 171–178.

Hambly, H., 1996. Grassroots indicators: measuring and monitoring environmental change at the local level. Leisa Mag. 12 (3), 14–15.

Hammond, P.M., 1994. Practical approaches to the estimation of the extent of biodiversity in speciose groups. Philos. Trans. R. Soc. Lond. B 345, 119–136.

Ho, P.H., 2000. In: Cay Co Viet Nam. An illustrated Flora of Vietnam, 3 vols. Youth Publishing House, Ho Chi Minh City (in Vietnamese).

Hong, N.X., 2002. Experiences of Human-Ecology Management in Ethnic Groups Ta Oi, Katu, and Bru-Van Kieu in Thua Thien Hue Province. Ethnic Culture Publishing House of Vietnam, Hanoi (in Vietnamese).

Howard, J.L., 2005. *Imperata brasiliensis, I. cylindrica*. In: Fire Effects Information System, [Online]. U.S. Department of Agriculture, Forest Service, Rocky Mountain Research Station, Fire Sciences Laboratory (Producer). Available: http://www.fs.fed.us/database/feis/ (28.11.06).

Hy, H., 2005. 'Cay Uoi bay' (*Scaphium lychnophorum*). In: Newsletter No. 5 of Science and Technology of Thua Thien Hue Province (in Vietnamese).

Institute of Ecology and Biological Resources (IEBR), Center for Resources and Environmental Studies (CRES), 2003. In: Danh Luc Cac Loai Thuc Vat Viet Nam (The Plant Name Index of Vietnam), 3 vols. Agricultural Publishing House, Hanoi.

Kati, V., 2003. Testing the value of six taxonomic groups as biodiversity indicators at a local scale. Conserv. Biol. 18, 667–675.

Kremen, C., Raymond, I., Lance, K., 1998. An interdisciplinary tool for monitoring conservation impacts in Madagascar. Conserv. Biol. 12, 549–563.

Lan, L.V., Ziegler, S., Grever, T., 2002. Utilization of Forest Products and Environmental Services in Bach Ma National Park, Vietnam. Online: http://www.mekong-protected-areas.org/vietnam/docs/bach_ma_forest_products.pdf (accessed 28.10.06).

Matavele, J., Habib, M., 2000. Ethnobotany in Cabo Delgado, Mozambique: use of medicinal plants. Environ. Dev. Sustain. 2, 227–234.

McCune, B., Grace, J.B., 2002. Analysis of Ecological Communities. MjM Software Design, Oregon, USA. pp. 198–204.

McCune, B., Mefford, M.J., 1999. PC-ORD. Multivariate Analysis of Ecological Data, Version 4, MjM Software Design, Oregon, USA.

McGeoch, M.A., 1998. The selection, testing and application of terrestrial insects as bio-indicators. Biol. Rev. 73, 181–201.

McLachlan, S.M., Bazely, D.R., 2001. Recovery patterns of understory herbs and their use as indicators of deciduous forest regeneration. Conserv. Biol. 15 (1), 98–110.

Ministry of Agriculture and Rural Development (MARD), 2000. Ten Cay Rung Viet Nam (The Names of Forest Plants in Vietnam). Agricultural Publishing House, Hanoi. 464 p. (in Vietnamese).

Ministry of Forestry (MOF), 1984. Quy pham thiet ke kinh doanh rung (Procedures for Forest Business Designing) (QPN6-84) issued by the Minister of Forestry on 1 August 1984 under Decision No. 0821B/QDKT. (in Vietnamese).

Moller, H., Berkes, F., Lyver, P.O., Kislalioglu, M., 2004. Combining science and traditional ecological knowledge: monitoring populations for co-management. Ecol. Soc. 9 (3), 2 [online]. http://www.ecologyandsociety.org/vol9/iss3/art2/.

National Research Council (NRC), 1995a. Review of EPA's Environmental Monitoring and Assessment Program: Overall Evaluation. National Academy Press, Washington, D.C.

Noss, R.F., 1990. Indicators for monitoring biodiversity: a hierarchical approach. Conserv. Biol. 4, 355–364.

Noss, R.F., 1999. Assessing and monitoring forest biodiversity: a suggested framework and indicators. For. Ecol. Manage. 115, 135–146.

Ostrom, E., et al., 2004. International Forestry Resources and Institutions (IFRI) Research Program: Field Manual. Version 12, Center for the Study of Institutions, Population, and Environmental Change, Indiana University, USA.

Rundel, P.W., 1999. Forest habitat and flora in Lao PDR,cambodia and Vietnam. Conservation Priorities in Indochina. World Wildlife Fund, Indochina Programme Office, Hanoi, Vietnam.

Salam, M.A., Noguchi, T., Pothitan, R., 2006. Community forest management in Thailand: current situation and dynamics in the context of sustainable development. New For. 31, 273–291.

Schmidt, I., Zerbe, S., Betzin, J., Weckesser, M., 2006. An approach to the identification of indicators for forest biodiversity – the solling mountains (NW Germany) as an example. Restor. Ecol. 14, 123–136.

Schmidt-Vogt, D., 1999. Swidden Farming and Fallow Vegetation in Northern Thailand. Geoecological Research, vol. 8. Franz Steiner Verlag, Stuttgart.

Sheil, D., 2001. Conservation and biodiversity monitoring in the tropics: realities, priorities, and distractions. Conserv. Biol. 15, 1179–1182.

Sikor, T., Apel, U., 1998. The Possibilities for Community Forestry in Vietnam. Asia Forest Network Working Paper Series. Asia Forest Network, Santa Barbara, California, USA.

Slik, J.W.F., KeBler, P.J.A., Van Welzen, P.C., 2003. *Macaranga* and *Mallogus* species (Euphorbiaceae) as indicators for disturbance in the mixed lowland dipterocarp forest of East Kalimantan (Indonesia). Ecol. Indic. 2, 311–324.

Steinmetz, R., Chutipong, W., Seuaturien, N., 2006. Collaborating to conserve large mammals in Southeast Asia. Conserv. Biol. 20 (5), 1391–1401.

Steven, D.D., 1994. Tropical tree seedling dynamics: recruitment patterns and their population consequences for three canopy species in Panama. J. Trop. Ecol. 10, 369–383.

Stork, N.E., 1994. Inventories of biodiversity: more than a question of numbers. In: Systematics and Conseravtion Evaluation. Claredon Press, Oxford, pp. 81–100.

Swamy, P.S., Sundarapandian, S.M., Chandrasekar, P., Chandrasekaran, S., 2000. Plant species diversity and tree population structure of a humid tropical forest in Tamil Nadu, India. Biodivers. Conserv. 9, 1643–1669.

Thang, T.N., 2004. Forest Use Pattern and Forest Dependency in Nam Dong District, Thua Thien Hue Province, Vietnam. M.Sc. thesis, Asian Institute of Technology, Bangkok, Thailand.

Thiha, Webb, E.L., Honda, K., 2007. Biophysical and policy drivers of landscape change in a central Vietnamese district. Environ. Conserv. 34 (2), 164–172.

Tordoff, A., Timmins, R., Smith, R., Ky Vinh, M., 2003. A Biological Assessment of the Central Truong Son Landscape. Central Truong Son Initiative. Report No. 1, WWF Indochina, Hanoi, Vietnam.

Whittaker, R.H., 1965. Dominance and diversity in land plant communities. Science 147, 250–260.

Assessing Conditions for Effective Community Forest Management in Thua Thien Hue Province

N.T. Duc*, L. Van An*, D.T. Duong*, N.X. Hong[†]

*Hue University of Agriculture and Forestry, Hue City, Vietnam †Hue University of Science, Hue, Vietnam

10.1 INTRODUCTION

Community forest management (CFM) is one of the models of social forestry or community forestry, and is the model that has been noticed and incorporated in the forestry development strategy of many countries around the world. In this model, local people have direct responsibilities for the activities of managing and protecting forests and receive the benefits for their contributions to the operation (Donald et al., 1993). In Asia, due to a general belief that it can improve the lives and livelihoods of about 450 million people living in and nearby forests, CFM has attracted special attention (Mahanty et al., 2007). It is shown through the models that have been implemented such as the model of forest management by household groups in Nepal, the cooperative forest management model in India, the model of forest management by villages in Indonesia and the Philippines, the model of forest management by villages and household groups in Viet Nam, and others.

In Viet Nam, the forest strategy of shifting from state centralized management to socialized forestry has been shaped and gradually been implemented since the 1990s. Especially in the early 2000s, along with the development of the institutional framework on CFM and related policies, CFM has become an official model in forest resource management (Nguyen et al., 2005, 2008). The expectations of the program and the model related to CFM are to practically contribute to sustainable national forest resource management, as well as contribute to improve livelihoods and reduce poverty for the local people (Bao, 2006). Thus, the opportunity for participation by local people as well as the rights, responsibilities, and benefits in the process of forest resource management and use have been considered and encouraged. As a result, up to Dec. 2007, 10,006 village communities were managing and using 2,729,946.3 ha of forests and bare land, in which 1,916,169.2 ha are land with forests (Nguyen, 2009a). Indeed, many CFM models have been tested and applied in the whole country with expectations and recognition of the linkage between a community's participation and poverty reduction issues in forest resource management. However, while the policies, institutions, and approaches used to develop this model still continue to be developed and improved, the specific practical aspects of policy implementation and confirmation of performance still need further study and evaluation (Nguyen, 2009a; Vo, 2010).

In Thua Thien Hue Province, the model of CFM has been applied on trial since 2000. At the time Thuy Yen Thuong Village, Loc Thuy Commune, Phu Loc District was chosen to initiate this model, around 400 ha of natural forests were allocated to the village community. Up to 2009, there were 10,904.7 ha of natural forests allocated to communities (10,327.5 ha) and households (577.2 ha) for long-term management in the duration of 50 years in the area of 4 districts, namely, Phu Loc, Nam Dong, A Luoia, and Phong Dien districts (TTHDARD, 2009). However, the problem of how to effectively manage the community forest is an issue that needs to be further experimented with and studied (Nguyen, 2009b, 2010).

Based on the practice of CFM mentioned above, this research is implemented in Thua Thien Hue Province with the specific objectives of (1) generally analyzing the characteristics of CFM within the study area, (2) analyzing the potentials and constraints related to the current forest management, and (3) identifying and proposing practical lessons to consolidate and orient the effectiveness of CFM. At the same time, through its findings, the research hopes to contribute to providing information for communities that applied and will apply the model of CFM, particularly the policy makers in terms of consolidating and adjusting suitable policies and institutions.

10.2 RESEARCH METHODOLOGY

10.2.1 Orienting Conceptual Framework for the Research

Based on the practice of research history and data sources of the survey area, this study is oriented to apply the assessment method related with a multiuse analysis framework in association with contextual factors rather than based on the development and application of the assessment through a set of criteria on effective CFM. The contents and process of the research are generalized and described in Fig. 10.1. This analysis framework consists of seven contents that can be summarized into three components: an overview and characteristics of the community forest allocation program in Thua Thien Hue Province; an analysis of the characteristics and other factors that bring the success of identified typical forest management models; and a summary of the lessons learned and policies oriented to contribute to strengthening the work of constructing an effective CFM model.

10.2.2 Study Location and Methods of Collecting and Analyzing Data

The model of CFM in Thua Thien Hue Province has been studied recently by some authors. The studies focus on analyzing aspects relating to institutions, management status, as well

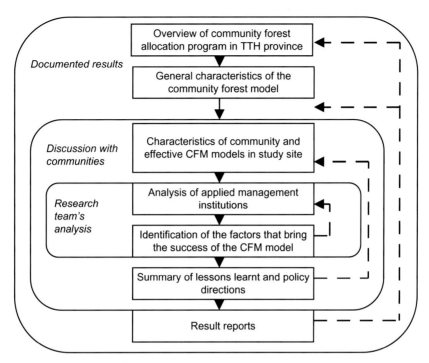

FIG. 10.1 Conceptual framework for the research.

as official and unofficial tenure rights of the local community. Based on the results of those completed studies, especially based on the oriented goals of the research, the study sites and data collecting methods are determined by a number of theoretical points as discussed below.

To study the natural forest allocation program for the community in the area of Thua Thien Hue Province, reports from agencies such as the forest protection department, the unit of forest protection, and the department of agriculture and rural development at the district level are summarized and analyzed. Particularly through the study reports on policies and current status of CFM form Ngo et al. (2009, 2013) and the study report on the current status and rights in practical CFM form Le et al. (2010), the characteristics and basic factors that influence the orientation of effective CFM are summarized and analyzed. These study reports are implemented based on the model of CFM in Hong Trung, Bac Son, Hong Kim, Hong Ha, and Thuong Quang communes (as shown in Fig. 10.2).

Based on the analysis report of secondary data, the models of typical effective CFM are determined to carry out detailed research and surveys (Fig. 10.2). Two typical CFM models, which have been chosen in this study case, are the model of forest management by Thuy Yen Thuong Village, Loc Thuy Commune (Nguyen, 2005; Nguyen, 2010), and the model of forest management by the interest group or club in Phu Mau Village, Huong Phu Commune (Pham and Duong, 2010). Through group discussions with CFM boards and key farmers, institutions and practices for implementing activities of these two models were collected and synthesized.

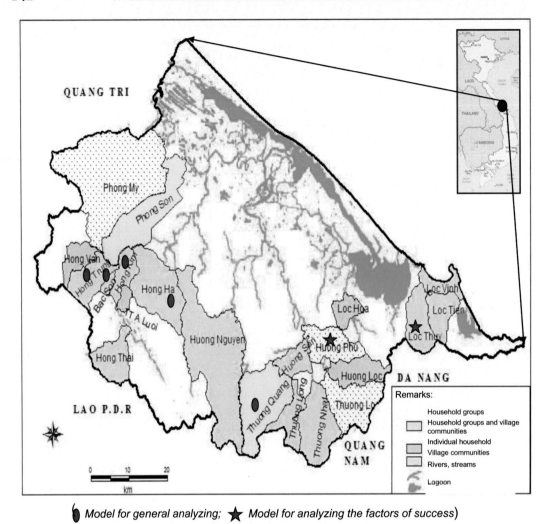

(⬤ *Model for general analyzing;* ★ *Model for analyzing the factors of success*)

FIG. 10.2   Forest allocation program implemented areas and areas selected for study.

From a general analysis of the results and detailed information from typical successful models, the conditions and factors that lead to the success of the CFM models are summarized and analyzed. Analyzing contents and discussions is integrated in the process of presenting the results and findings of the research. Analysis methods applied throughout the research took the principles of theoretical studies, combined with the findings summarized through practical research documented to compare with the reality of the models used to analyze details in this study. The analyzed results were exchanged and discussed with the community and stakeholders to check the reliability and identify orienting solutions for effective CFM.

IV. MERGING SCIENCE AND TRADITIONAL PRACTICES IN NATURAL RESOURCE MANAGEMENT

10.3 STUDY RESULTS AND DISCUSSION

10.3.1 Overview of Forest Allocation Program for Communities in Thua Thien Hue Province

Thua Thien Hue Province in the whole country taking the lead on approaching the model of allocating natural forest for communities. From the time of initiating the model in Thuy Yen Thuong Village, Loc Thuy Commune, in 2000 under the support of the Program on Forests (PROFOR) project, 10,904.7 ha of natural forest had been transferred for communities and households to manage in the long term as of Dec. 2009, in which the area allocated to the community was 10,327.5 ha (TTHDARD, 2009). The allocated areas corresponding to each district are shown in detail in Fig. 10.3.

Natural forest allocation programs for communities to manage are applied primarily in four of eight districts in the province. The allocated area of forest in A Luoi District has the highest proportion (57.6%), followed by Nam Dong District (21.5%), and the lowest proportion is Phong Dien District (6.6%). Types of allocated forest are mainly protective forests and production forests, and in Nam Dong District only production forests were allocated up to the present. By analyzing the current status of forest allocation in two typical districts, A Luoi District and Nam Dong District, some general characteristics of the community forest allocation program have been summarized and shown in Table 10.1. These are two mountainous districts located in the riverhead of two main rivers (Huong River and Bo River) of Thua Thien Hue Province, where the natural forest allocation program has been implemented since 2003. The total natural forests of these two districts occupy around 64% (146,279 ha) of the province's total natural forest area (VNFPD, 2008; ALFPU, 2008; NDDPC, 2009).

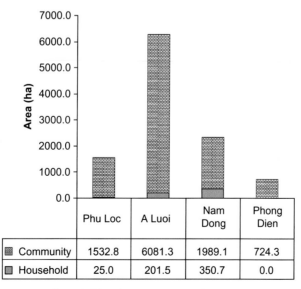

	Phu Loc	A Luoi	Nam Dong	Phong Dien
Community	1532.8	6081.3	1989.1	724.3
Household	25.0	201.5	350.7	0.0

FIG. 10.3 Allocated forest area to household and communities in the province.

TABLE 10.1 Characteristics of Forest Allocation Program in A Luoi and Nam Dong Districts

District	Total Allocated Area (ha)	Type of Forest (%)		Type of Allocation (%)			Supporter and Supported Area (%)				
		Production	Protection	Individual Household	Household Group	Village Community	SNV	ETSP	UNDP	GCP	FPD
A Luoi	6283.4	1.6	98.4	3.2	–	96.8	79.1	6.3	9.4	1.0	3.2
Nam Dong	2478.6	100.0	–	14.0	58.0	28.0	51.2	2.4	–	2.4	44.0

SNV, Stichting Nederlandse Vrijwilligers (Netherlands Development Organization); *ETSP*, Extension Training Support Program; *UNDP*, United Nations Development Program; *GCP*, Green Corridor Project under WWF; *FPD*, Forest Protection Department of Thua Thien Hue Province.
From ALFPU (A Luoi Forest Protection Unit), 2008. Report on Assessment of Forest Allocation Program Within the District; NDDARD (Nam Dong Department of Agriculture and Rural Development), 2008. Report on Land Use Planning and Forest Land Allocation.

Table 10.1 shows that the natural forests are allocated to local people under three types: household (hh) type, household group type, and village community type, in which the model of allocating the forest for village community takes the highest percentage. These models are implemented under different methods of forest resource assessment to construct benefit mechanisms (reserves and stable forest models), and under technical and financial assistance from different organizations. Supporting organizations are the Netherlands Development Organization (SNV), the Extension and Training Support Project (ETSP), the United Nations Development Program (UNDP), the Global Canopy Program (GCP), and the Forest Protection Department. Of these five organizations, the SNV plays the most important role.

To understand the basic characteristics of the CFM model in Thua Thien Hue Province, Thuong Quang Commune, Nam Dong District, Bac Son Commune, and Hong Kim Commune in A Luoi District have been chosen for analysis. In Thuong Quang Commune, the model of CFM exists in two types: the household group type and the village community type. The household group model was implemented in 2003 under the support of SNV and the village community model was implemented in 2004 under the support of ETSP. In Hong Kim and Bac Son communes, the applied model is the model of forest management by the community, which was implemented in Bac Son Commune in 2006 and in Hong Kim Commune in 2007. The analyzed results of these models are shown in Table 10.2.

The analyzed results from Tables 10.1 and 10.2 show that ever since the moment the forest allocation program was initiated, the model of CFM has been implemented under different types and conditions. That is reflected in the types and area of allocated forests, user groups, resource assessment ways to construct the benefit mechanism, as well as the organizations that offer financial and technical support. Major allocated forest types are protective and production forests, and are mostly concentrated in the status of medium and poor forests. An average allocated area toward the model of forest management by household group is 75.8 ha and by village community is 98.5 ha. The number of households in the model of forest management by household group varies from 6 to 10 households, and 40–70 households in the model of forest management by village communities. Each household group or village community sets up forest management teams with the number of households varying from 2 to 10 depending on each circumstance. As regards the household group model, the number of members in one team varies from 2 to 4; in village community model it is 5–6 households in the case in Hong Kim Commune and 8–10 households in the case in Thuong Quang and Bac Son communes. In both models, not all people within the commune are participating in the activities related to CFM, from which around 24% of people participated in the household group model, and around 68% in the village community model.

10.3.2 Characteristics of the Models of CFM Related to Tenure Rights

Allocating the natural forest for the community to manage and use in the long term is one among other types of legal tenure right allocation for the community, which is an important basis for the community to trust and confidently invest in the process of managing and using the forest resources, as well as legally and sustainably protecting their tenure rights when events occur. In the model of CFM in Vietnam, the forest receiving community was given four basic rights in the bundle of rights theory discussed and analyzed by Schlager and Ostrom (1992): the right of access, exploitation, management, and exclusion, in which, the right of

TABLE 10.2 Typical Characteristics of CFM Models in Three Surveyed Communes

| Type of Allocation | Allocated Area (ha) | Model-Based (%) | | Participants per Commune (%) | Average Number of hh | Average Forest Area per Group or Village (ha) | Average Forest Area per hh (ha) | Type of Forest[a] (%) | | |
		Forest Reserves Model (FRM)	Stable Forest Model (SFM)					Medium	Poor	Restoring
Group of household	909.3	100	–	23.6	7[b]	75.8	10.9	74.2	17.5	8.3
Village community	1045.1	56.4	43.6	67.6	56[c]	98.5	1.8	38.9	53.1	8.0

Poor and restoring forests: timber reserves are about <80 m³/ha.
[a]*Medium forest: timber reserves are about 80–150 m³/ha.*
[b]*Number of households varied from 6 to 10 households.*
[c]*Number of households varied from 40 to 70 households.*
Data from Author's survey, 2009

exploiting the valuable products, such as timber, was specified more strictly and tightly than other products. The right of converting the purpose and type of resource usage in management rights is limited. These tenure rights are specified in the institutions and policies such as the revised Land Law 2003, the revised Forest Protection and Development Law 2004, the revised Civil Law 2005, Decision 178/2001/QĐ-TTg, Decision 186/2006/QĐ-TTG, Decision 106/2006/QĐ-TT, and so on. However, as a practical reality, each CFM model has its own characteristics related to specific community and forest resources.

In the CFM model, legal tenure and illegal tenure always exist in parallel, which creates supporting, overlapping, and even contradicting issues. Through the analytic reports of Ngo et al. (2009, 2010) and Le et al. (2010) on some CFM models in Aluoi and Nam Dong districts, several issues related to reality application of institutions and tenure rights have been detected. It is summarized and presented in Table 10.3 and in the detailed analysis below.

TABLE 10.3 The Basic Issues Related to Implementation Status of CFM Model

No.	Category	Implementation Status
1.1	Delivering documents of appropriate authorities	Slow process; lack of detailed documents on the rights and benefit mechanisms in case of forest allocation to household groups
1.2	Community using documents	The forest management board is passive in understanding and propaganda. Community members do not have any documents related to their forests
2.1	Setting up organizational structures	Implementation is fairly synchronous; however, it is lack of legal decision for forest protection team; the models of forest management by village communities are still formalistic heavily
2.2	Constructing forest protection and management convention	The role of members and regulations is clearly defined. However, it is lack of detailed information on sharing benefits within community
2.3	Building patrol schedule	The plans are built in detail on implementation time and objects
2.4	Propaganda	Organization form of community is rather monotonous; forestry agencies involved in implementing these activities irregular
2.5	Detecting and handling violations	The violations are mostly within community; violations handling has not been effectively implemented as mentioned in the convention
2.6	Educating violations	Community has not implemented or implemented effectively on educating people in the community when they violate the regulations
3.1	Numbers of participant	3–5 persons in forest protection team; forest management boards are mainly implementing this activity in comparison with other members in community
3.2	Implementing patrols	This activity is implemented quite well in the first phase; the rules have been broken after 1 year of implementation for most of forest management model by village community
4.1	Forest resource using and community awareness	Timber resource and NTFPs play an important role in community livelihoods; communities have not effectively used the resource using rights which have been established
4.2	Nurturing forests	Communities are not interested and lack of technical knowledge

Continued

TABLE 10.3　The Basic Issues Related to Implementation Status of CFM Model—cont'd

No.	Category	Implementation Status
4.3	Artificial regeneration	It is formalistic and dependent on the investors (outsiders)
5.1	Training on forest management	Appropriate authorities have periodically deployed 1–2 times per year; however, it still has awareness propaganda nature than guiding basic techniques
5.2	Guiding documents on handling violations	Communities are not equipped documents and knowledge related to constructing treatment records of violations to ensure legality
5.3	Guiding documents on silvicultural technique	Community is not equipped these documents
5.4	Constructing fund for forest protection, management, and development	Communities are passive in implementing this activity; appropriate authorities have not shown promoting and supporting role to promote community internal force
5.5	Developing livelihood models	There are no livelihood models which are developed based on forest resource and forest land; lack of orientation and investment of stakeholders
5.6	Stakeholders' participation	Not mentioned clearly in the convention; the comanagement role has not been shown clearly

10.3.2.1 Rights of Forest Resources Access and Exploitation

Forest resource accessing and exploitation rights and enjoyed interests are defined in the policy institutions of government and forest protection convention of the community. These rights are expressed in many diversified aspects of the resource. The benefits of exploiting timber products, nontimber forest products (NTFPs), landscape ecology, farmland, and other mineral resources. Actually, the communities are allocated forests that are usually ethnic minority communities, located in remote areas, and have livelihoods dependent on forest resources. Within these communities, forest resource access and exploitation activities management has been established and run before the forests are legally allocated to the community. At the points of demarcation, appropriated rights were captured and recognized by the members of the community. Boundaries are only based on the local terrain and natural objects or making the resource such as cultivation and "ken" of trees to whom assert resources belonged to. The viewpoint: "forest is the resource that nature has bestowed so that it does not belong to own one and anyone can approach and exploit" still exists in local people's awareness. This has created difficulties and obstacles for the forest management and protection work of the community when they are empowered legally. Can Sam village community has met difficulties when restricting illegal access and exploitation of the members of other communities.

Conflicting views about resource exploitation rights have been shown quite clearly in the communities. When being allocated the self-control rights on forest management and protection, illegal exploitation people groups in Hong Kim commune community have hastened resource exploitation through using the sawmill to exploit timbers, "ken" of trees to regularize which can be exploited, and so forth. Illegal timber exploitation by people in the community is also happening in other communities that are allocated forests such as Can Sam Village (6.9% of household in the village) and Le Trieng Village (3.2 % of household in the village).

The community's promotion and usage right in land use for cultivation and timber exploitation are still limited. The shifting land and the grassland are permitted to be used have not been planned or used effectively and spontaneously by the community. In most forest areas allocated to the community, the area of grassland and weak recovery forests has always existed. For example, Can Sam Village has 20 ha; Hong Kim Commune has 50.5 ha; and Hong Trung Commune has 40 ha. However, these areas have not been used effectively to create a livelihood source for the community by implementing agroforestry cultivation models, production forest cultivation, and so on. Particularly, illegal transgressions of some community members have happened on this land. For example, Mr. Ho Van Tinh and Mrs. Ho Thi Khoa of Can Sam Village have transgressed 3 ha of community forests for planting fast-growth forest. Mr. Ho Van Vang and Mr. Ho Van Phuong of Hong Kim Commune have transgressed 20 ha of community forest for acacia planting. This has created dissatisfaction and contradiction among community members. Related to exploiting timber that is differential of the reality forest model and stable forest model, communities have not implemented this as part of the forest management plan that is established by them.

10.3.2.2 Forest Management and Protection
ACTIVITIES ON BUILDING AN ORGANIZATION SYSTEM AND ENHANCING AWARENESS

During the period of building the procedure for forest allocation and after forest allocation, the organization system of forest management, protection, and development was established completely in the community. The village forest management board and forest protection team were set up and got the requirements of quality, as well as the regulations about roles and responsibilities were also expressed in details. However, there were some limitations in approaching and handling the information, as well as the difficulties of the economy, labor, health, and age of some villagers, so they were limited in participation in forestry management and protection activities. In addition, the difference of distance to forests and forest resources situation among the communities has made a difference in the number of actual members who participated in these activities. In hamlet 1, 2, and 3 of Hong Kim Commune, and hamlet 1 and 3 of Bac Son Commune only 50% of the community participated in forest management and development activities.

Propaganda to raise awareness about the policies and the importance of forestry management and protection has been conducted by the CFM board. The activities were integrated in the village meetings, or local forestry agencies meetings held periodically twice per year. However, the training activities to improve the abilities of forest management, protection, and development still have some problems. In the process of allocation, there are only some people who are in management, protection, and development on the community forest board of the village or the group that joined these activities, and the key farmers who participated include only about 10 people and they only joined only in the first steps in the process of allocation. After the allocation, meetings and training just focused on the improvement of awareness more than equipping the skills of forest management, protection, and development in the community. Especially in these meetings, people who often impact on forests did not participate. Additionally, due to delays in the delivery of the needed guidance documents to the community from the related agencies (the communities in Hong Ha and Bac Son have not received the necessary documentation) or the community does not receive any relevant documents (model of households in Thuong Quang), the passive manner in researching documents (in Hong Kim, guidance documents are almost not used) are a cause of the limitation in implement and deal with the activities. Therefore, they cannot avoid the lack of understanding of policies as well as the incomprehensiveness in applying the policies.

THE PROTECTION PATROLS OF FOREST ACTIVITIES

Forest protection patrols must be conducted according to the agreement in the conventional forest management. The forest protection group of the community will patrol and observe the forest area. The groups also have to make a diagram of necessary patrol ways and important areas. The patrol was deployed quite well in the first period after receiving the forests. However the gap between the times increased to 2 months as in villages 1–3 of Hong Kim Commune, and even 6 months in Bac Son Commune (did not do the patrol from Jan. to Jun. in 2008). The reason for this problem, as explained by the people, was the lack of fund to pay for the patrol as well as the legal decision for the forest protection team, so they have no motivation to do their tasks. Although it has been clearly mentioned in the policies and rules for protecting the community forest that the community will get the benefits through their activities, over 90% of participants wanted to receive a salary directly after their patrol participation. Additionally, the patrol activity only followed a fixed route so this has limited the effectiveness of the patrol work.

Illegal forest exploitation and the complexity and sophistication of the "culprit" illegal forest invasion are important factors affecting the effectiveness of protection activities. According to regulations, people can exploit the products from dried trees, dead trees, and so forth. Thus, there were many cases of deliberate impact to make trees die before exploiting (trees "ken"). This activity is becoming quite popular in the area of Bac Son Commune and Hong Kim Commune. Trees with 20–30 cm diameters in forests located at the junction between the communes (100–200 m from the boundary) are often clearly compromised. According to statistics of the community, 34 violations were discovered in the research area in the year 2009, of which 29 cases were in the village community model and 5 cases were in the household group model. The important thing to note is that 80% of the cases caused by the violations belong to the community. However, the handling of violations based on the forest convention has not been done thoroughly, and only 30% of the violations were handled. Emotional problems, relationships, economic status, equitable reason of the violation, as well as awareness of the forest management board are the root cause of all the unhandled situations.

Forest patrol activities involve hard, physical work, so men are the main participants in these activities. But some people who have been assigned to forest protection were not guaranteed to perform it. In most of the cases there were at least one to two people absent for many different reasons and this did not exclude legitimate reasons. The irregular work and the boundaries separation assigned to patrol between the communities were shown to be a reality. For example, village 1 and village 2 in Bac Son and villages 1–3 in Hong Kim have done this irregular work more often than other villages.

FOREST DEVELOPMENT ACTIVITIES

Forest development activities are an important activity in addition to forest protection activities. These activities can take place in many different forms, such as zoning regeneration, nurturing care, regeneration, and artificial reforestation. However, due to poor economic conditions, depending on natural resource exploitation rather than investment and the impact of this on forests has made this activity unlikely to be done here, or done with the support of organizations. Hamlet A Ro has been planting 100 rattan trees under the shape of forest with the support from the Green Corridor Project, group 2 of village 6 in Thuong Quang planted 1 ha of indigenous forest trees with support from the Consultative and Research Center on Natural Resources Management (CORENARM), and planting indigenous trees model under forest shade and the community nursery model supported by the Green Corridor Project to Can Sam Village. Additionally, pressure from the economic model of fast-growth forest (acacia) created an increased risk of illegal land use for application of this model. Guidance documents of simple sylviculture techniques that should be used to help the community to implement the care and nurturing of forests still did not equip the community.

10.3.2.3 Community's Benefits and Livelihood

The overall benefits of community forests for the livelihoods of local people have many aspects, such as direct benefits from forest products and benefits of ecoenvironmental values, social benefits, and so on. Scale, level, and impact of the direct or indirect benefits will vary between different objects in the community. It is worth noting that how the community can recognize the overall benefit and nonmaterial benefits that the forest resources have been giving them is a subject that should be considered in the forest management model. In fact, local people focus their attention mainly on direct benefits from forest products, especially wood, and in the present rather than for the future. One of the reasons is poor economic conditions and insufficient awareness about the overall value of the forest resources of communities who received forest. In the community where forests were allocated, the proportion of poor households is high, such as 23.6% in Thuong Quang, 61.8% in Hong Kim, and 62% in Bac Son (2007).

Harvesting NTFPs is an activity that is quite typical of mountain people in general and particularly in the research area. NTFPs have been used as food, construction materials, and production material for the family; they also contribute to the generation of additional income for some families when a product is sold in the market. The average level of contribution of forest products for livelihood was about 15–20% of households in which poor households still show obvious dependence on this product. These products can play a role as fuel (A Ro Village and village 6 in Thuong Quang exploited about 50–100 tons of firewood from the forest per year and a household in Hong Kim exploited about 1 m^3 per month of firewood from the forest), and also as materials for building houses, sheds, and fences; all of which are clearer than the food supply. The change in habits and frequency of exploitation did not show clear differences between before and after the forest allocation.

The policy in supporting timber in the period before receiving the benefits of timber products due to growth of forests has also been applied in the communities that received forests. After receiving forests there were 9 households in Bac Son Commune and 14 households in Hong Kim Commune that were supported with the timber to build houses; A Rong Village and village 6 in Thuong Quang Commune were allowed to exploit 23 and 10 m^3 to build the house. This activity was done before the forest allocation, but it has become more important since the forest was allocated to the community for protection and management. Benefits related to wood products are due to the growth of allocated forests, and the reserves of forest are low now (for the reserves model) and the number of large-diameter trees is limited (stable forest model) so the near-term benefit is limited. Based on the average growth time of forests, it takes 10–15 years to set exploitation standards and for technologies to mature. So this will be a barrier for communities to pay the costs related to protection and management activities. Even cases based on the stable forest community model are confusing, and this has not helped in the exploitation of trees that can be exploited in different diameter sizes.

10.3.3 The Reality of a Typical Model of Effective CFM

10.3.3.1 Model of Forest Management in Thuy Yen Thuong Village

Thuy Yen Thuong Village belongs to Loc Thuy Commune, the distance to the center of Phu Loc District is about 13 km to the southwest, and the transportation system and communication are convenient. This village is located to the basin of Suoi Tien. This stream originates in mountainous areas of Bach Ma National Park, and is also an important source of water supply for daily life and production of the villagers. Natural forest areas located at the top of the village's hills play a quite important role for the livelihoods of 395 households with 1900 inhabitants (2009). The history of people's impact on forest resources, changing awareness processes, the opportunities in receiving forests for management and use, as well as the reality of implementation of management and protection activities of the community, is a process full of difficulties and challenges, but also sources of pride for the achievements already made. Essential progresses and achievements are demonstrated through several points discussed later.

CHARACTERISTICS OF LIVELIHOODS AND CHANGES IN THE AWARENESS OF THE COMMUNITY PRIOR TO THE ALLOCATED FOREST

The community in Thuy Yen Thuong is a Kinh ethnic community who have lived here for a long time. During the war period, many households in the village were evacuated. After the country completely unified in 1976, people returned home to start building a new life. At this period in agricultural production, paddy rice cultivation is the main production and livelihood source of the people here. However, it still cannot adapt to the basic livelihood needs of the community. Illegal exploitation of forest resources is a "solution" the local people have selected. Although the forestry agencies and local governments have tried to stop it, the exploitation of forest products, especially wood, went massively out of control in many areas of natural forest within the village. Sometimes there were 50–60 buffaloes in the village to carry the wood. This illegal activity continued until 1997 and 1998.

Facing this situation, in 1998 a Thuy An forest protection station was set up and stayed at the forest gate of the village, and at the same time Thuy Yen Thuong Village was legally formed. The forest protection station was formed for the purposes of limiting and preventing the illegal action of communities on forest resources, supporting and advising on techniques, and offering legal advice and raising awareness of forest protection for the community. The community has been invested in guided techniques in sowing seed and planting forests, provided more chances of getting jobs through forest care activities, and tending and protecting forests (about 400 employees per quarter). In line with the integrated activities of the local forestry agencies, improving the awareness about the role of forests and the importance of forest management in the community began at first with the older and key members in the village. Forest resources have declined severely and people cannot live forever with these illegal activities. The thoughts and concerns of the villagers are that their children will suffer the consequences. This thought has been shared and discussed in community meetings, and then they decided to receive forests to protect, manage, and use. Their proposal was considered by the district forest protection unit and local government. During the period from 1998 to 2000, the community showed the evidence by limiting and reducing the illegal activities in the use of forest resources. This was also the evidence to secure the trust of forestry agencies and local government. Finally the wishes of the community have been satisfied; they agreed to build the first model for CFM in Thua Thien Hue Province.

At the time this experiment of the forest management model was considered, the community had 381 households with 1860 people, and agricultural labor accounted for 97% of 856 employees (LTCPC, 2000). This is one of the poorest villages of Phu Loc District, with approximately 590 m^2 of agricultural land, average food of 200 kg/person/year, and income below 100,000 VND/person/year. Many of the difficulties that communities face in such situations were present: a low, intensive level of agricultural production, floods and droughts threatening every year, simple production (no mechanization) so the crop yields of were low, livestock that did not develop, no work to do after crops season, and a part of the community still exploited illegally the forest resources. However, awareness and desire to protect the forest formed as a consequence of the problems mentioned above, as well as the impact of the historical consequences of floods in 1999 and the formation of Thuy An agricultural cooperative (the existence of forests is critical for the tourism business in the Suoi Tien spring).

FOREST MANAGEMENT, PROTECTION, AND USE BY THE COMMUNITY AFTER ASSIGNING FOREST

Under the support and advice on the technical and financial aspects of the PROFOR project of the Ministry of Agriculture and Rural Development for the local government and concerned agencies, the community was allocated 405 ha of natural forests to protect and manage for long-term use with the process and basic legal procedures of a community forest allocation program. The construction of an organizational structure, rules of forest management and protection was also made within and after the allocation. A management board of forest protection including three members of village authority, and a forest protection team with 20 members who representatives of 10 inhabitants group (two people per group: leader and vice leader) were created. The regulations in the forest protection convention were drafted clearly and focused particularly on the necessary work allowed, encouraged to be done, or that cannot be done, as well as the rights and responsibilities of the community and each villager, and the regulations of rewarding and penalizing. These regulations were considered, checked, and corrected to be suitable two consecutive years after issuance.

Once the necessary steps and procedures as mentioned earlier were completed, the practical activities of forest management were applied quite flexibly and effectively. The forest management board and forest protection teams have identified that forest patrol activities are only an auxiliary solution for checking and evaluating forest resources rather than a permanent way to protect, prevent, and reduce illegal forest exploitation. Therefore, they have stepped up promotional activities to raise awareness of forest protection for people as well as the promotion of career-changing efforts and production development. Through village meetings where they organized by themselves, the management board will integrate the propaganda activities. Not only that, through effective agricultural production and legal economic activities that are compared with the previously illegal impacts to the forest, they have adopted propaganda and mobilized their colleagues to reduce the illegal exploitation activities to forests and seek suitable livelihood activities. The propaganda is not limited to formal meetings but also includes mobilization through community activities and events such as weddings, anniversaries, festivals, family or friends' parties, and so on. With the synchronized efforts of all members of the protection management board and the collaboration with the elders, respected people in the community, forestry agencies, and local government, the awareness and action to protect the forest of the community has been markedly improved. Until now, only few cases of illegal impacts to forests occurred in the community.

Initially, the community got the achievement due to the right attitude and the leadership apparatus capable of a forest management and protection board. However, the attention and coordination of forestry agencies and local government, and the policies in creating the opportunities to get the suitable benefit were the driving forces and the important lever that contributed to this success. In propaganda activities and forest protection patrols, the Thuy An forest protection station, forestry division, and farmers association of the commune have always been cooperating with the community. The food assistance during patrols and the advice and help in the policies information of the forest protection station created trust within the community. In addition, the policies to advance the community to get the timber from the forest that they are managing is a solution to solve the difficulties of financial and material resources the communities are facing.

With 405 ha of natural poor forests (average reserves: 76 m³/ha) allocated to the community to protect and manage, 90 m³ of timber in an area of 60.6 ha was agreed to exploit by concerned agencies in 2003. The exploitation activities were implemented and completed in 2004, with 78.8 m³ of timber (28 trees) exploited (52.8 m³ of timber brought back to the community). With this timber, the community made the sharing of benefits fairly and clearly (Fig. 10.4).

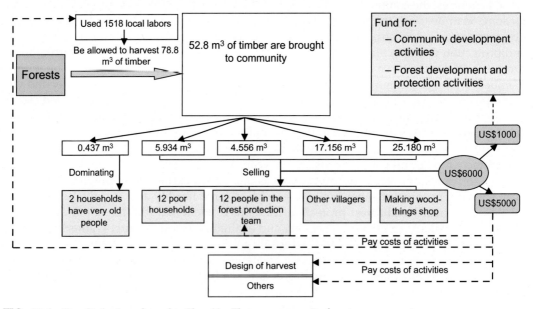

FIG. 10.4   Benefit sharing of wood in Thuy Yen Thuong community forest management.

Households in the village registered the purchase of wood through village meetings and they were approved under the agreed plan. Households outside the village who were beneficiaries of social welfare were also registered (with the opinions of the commune) if they had the requirement. The beneficiaries of social welfare had first priority, followed by households who contributed to the forest management and protection, and finally the woods workshops. There were $0.437\,m^3$ of timber support to households in need for making coffins; $5.934\,m^3$ of timber was sold to 12 poor households, solitary or well-deserved of the revolution; $4.556\,m^3$ of timber was sold to 12 households in the forest management and protection board; $17.165\,m^3$ of timber was sold to other people in the village; and $25.180\,m^3$ of timber was sold to the carpenter's workshop in the village and commune. In the total of 119.05 million VND (about US$6000 in 2010) from the sale of timber, the distribution of benefits to members who directly participated in timber exploitation, forest management, as well as the community was carefully considered. There were 88 workdays of forest management board and protection teams in 2003 and 2004, and 1518 workdays of people who exploited and transported timber were paid (protection and management: 25,000 VND/workday, exploitation and transportation production: 59,000 VND/workday). A part of the 19.2 million VND (about US$1000 in 2010) remaining amount was used for the construction of a community greeting gate; another part was used for management activities and the forest protection and development fund.

From the actual activities of forest management, protection, and use that have been made for the past 10 years, the community (especially the people on the management and protection board) have realized the appearance and effectiveness of activities that they implemented. People have responded and performed the responsible conventions of forest management and protection, the forest has been well-protected, and environment (especially the water source) has been improved. Policies that allow local people to get the benefit from timber products of the forest, combined with distributing the benefits equitably and clearly within the community, have promoted local people to believe in the forest allocation policy and actively participate in forest protection. The biggest lessons that the community learned were the need to create an effective combination of support and help from the agencies and the awareness and promotion of internal resources of the community, with the starting point being typical people in the community.

10.3.3.2 Forest Management Model of an Agricultural Extension Club in Phu Mau Village

A forest management model in the form of a club is an interest groups kind of management model. This model is formed firstly and uniquely in Thua Thien Hue Province at the current time. The characteristics of the model are not the fundamental reason leading to the attention of researchers or the locals that want to apply the model of CFM. The attraction was created by the "unique" procession in building and developing the management policies and the effectiveness that model brought in the process of forest resources protection with the orientation to develop community livelihoods. The general information about this forest management model is summarized and expressed through some points below.

FORMATION OF A CLUB AND OPPORTUNITIES OF RECEIVING FORESTS TO MANAGE, PROTECT, AND USE

Phu Mau Village is located on the northwest are of Huong Phu Commune, about 10 km from the center. The village was formed in 1994 by the resettlement program to create a new economic area. People came from Thuong Long Commune, Thuong Nhat Commune, Huong

Loc Commune, Khe Tre Town, and some delta areas. Currently, there are 92 households with 432 inhabitants; 22 households are ethnic minorities (mainly Ko Tu ethnic minority). Besides the area of natural forestland that has always existed, the area for agricultural and forestry production are about 400 ha, including 109 ha for short-term crops and long-term crops, 180 ha for rubber trees, 14.6 ha for paddy rice, and 150 ha is land for forestry production.

With the objective of creating a place to share experiences and help each other in the process of developing agricultural production, 25 households in Phu Mau Village proposed to form a group as a club. The aspirations of these households were agreed to by the district extension station, who consulted, supported, and promoted the decision of the People's Committee of Huong Phu commune about the establishment of the agriculture-forestry-aquaculture extension club of Phu Mau Village (referred as a club) in Oct. 2004. And since then, the regulations of the club were built and united among its members. In these regulations, rules and criteria for membership application, the functions and responsibilities of each member, activities schedule, financial use, award and punishment were described very clearly. It is worth noting that the provisions of the monthly fund of each member (5000 VND/month) included regulations for raising capital when needed for developing activities, regulations for payment from the income of activities (pay 20% for the club), as well as regulations to have remitted funds from the credit interest of 1% to the club's fund (supported 50% to capital and 50% to the club's fund). These are typical differences from the regulation of forest management and protection applied in most models of Thua Thien Hue Province, although the building fund of forest development and protection as well as mechanisms sharing the benefits have been assessed as playing a quite important role to maintain and mobilize the community's participation.

From the day of its establishment, the club realized its role in the protection and development activities of forest resources in the area. Currently, the forest resources (400 ha) are not owned by anyone in the area, but have been affected by people inside and outside the community. So, the club proposed an aspiration to receive a part of forest to manage, protect, and use based on the policies and institutions of the state. Thirty hectares of natural forest have been temporarily assigned by the local government and relevant authorities for the club to manage and protect as a step to test and evaluate the role of the club. Thus, the year 2004 marked the start of management, protection, and development of the forest by the club.

ACTIVITIES OF FOREST MANAGEMENT, PROTECTION, AND USE BY THE CLUB AFTER RECEIVING FORESTS

Forest management and protection activities for the first 2 years after the forest was allocated showed satisfactory results. The illegal exploitation activities over the forest area of the club were limited by coupling propagated activities in community meetings and periodical patrol activities of the club. A strong point is the members of the forest management board are also the leaders of the hamlet as well as the union (Mr. Duong Ca is the club's chairman and the hamlet's leader; Pham Doan is both vice-chairman and secretary of the hamlet and an ethnic minority), so forest protection propagation activity and assertion of sovereignty was developed advantageously. In addition, the way the club manages the forest of Club is mainstream, which brings local people to the agricultural cultivation and production area. So, other people in the hamlet contributed to monitoring and reporting if there are some disadvantageous impacts from activities in the forest.

From 2006 to 2008, under support and investment of the Swedish International Development Agency (SIDA) environmental fund (SEF) through Thua Thien Hue Nature

Care Union (Nature Care), the model of sustainable natural forest management and protection based on the right of people was built and carried out at the Club. Activities related to forest management, protection, and use had obvious advances, with high organization and planning. System consolidation was created and carried out first in this model. An organizing structure, style, and activity mechanism was rebuilt. Especially, activities will be carried out following the tendency to promote local sovereignty, responsibilities, and local knowledge in connection with scientific and technical applications, strengthening the conversation about activities to help members understand and carry out policies as well as the law of government to contribute to raising livelihoods and forest resources protection and development.

With those guidelines, a lot of activities to increase awareness and ability for members of the club to build a livelihood support model and implement forest resources development have been carried out. First is the collection and documentation of policy information related to rights, obligations, and duties of people about land resources and forest resource management and use (Land Law in 2003; Decree 181/2004/NĐ-CP, Decision 178 QĐ-TTg; Decision 186/2006/QĐ-TTg, and others). Information was edited and arranged clearly and was easy to understand and use for members in the club. Then the club organized conferences about policy understanding, forest enrichment methods, and so forth. In particular, the club organized training courses about the skills of planning, group working skills, as well as productive techniques for every production model. Up to now, the club has carried out five conferences and nine training courses for members in the club. Through conferences and training courses, local people had a chance to show their ideas, join the conversation, and ask questions about the problems they are facing. Members in the club promoted (1) awareness and knowledge related to rights and duties in forest management, protection, and development; and (2) knowledge about farming and breeding, care techniques, and raising and forest enrichment. From there they can make proposals and decisions to develop livelihoods and forest resources for their group.

Besides the patrol activity for forest protection that were carried out regularly by members of the club (each month a group took charge of forest patrol; each group had four to five members), models of forest resources and livelihood development were considered and taken seriously (Table 10.4). A planting and caring for rattan model, planting bamboo for bamboo shoot model, Lo O planting model, indigenous forest trees model, and economic forest model have been carried out within the allocated forest area. A porcupine raising model was carried out at the households level. Although most of these models (except the economic forest planting model) received support from government and nongovernmental organizations in the developing period and modeling investment, sustainable management has clearly slowed. After models were established and developed (application of participatory approach methods), devolving management mechanism through the tenders was applied forcibly. Households have aspirations in management and get benefits from this model put forward in the aspiration. The club carried out meetings to check and select households who have the best ability, and then a contract about management mechanisms and getting benefits properly was signed between the received household and the club. This method contributed to clearly define the management of each model; especially to promote a sense of initiative, duties, and authority for management. Besides the profit that the management subject got from the management and exploitation model, other members in the club also get benefits from extraction income of management subjects of this model (regulated clearly in the contract). This extraction income will be contributed to the activities' fund of the club. This fund will be used to develop

TABLE 10.4   Livelihood and Forest Resources Development Models at Club in Phu Mau Village

No.	Model	Characteristics
1	Rattan model	10 ha of rattan intercropped in forests (2004–05); care and protection of 20 ha of rattan in forests (2009): total support fund is about 60 million VND (Nature Care and CORENARM); 20 ha of rattan in forests will be assigned to a member of the club. Benefit sharing mechanism: 70% will belong to the household who manage model, 30% is given for the club
2	Bamboo for bamboo shoot model	2500 trees (1500 trees in forest, 1000 trees will be divided to households); this model was carried out in 2007 with support of agricultural extension station of district, model was assigned to one household in 2010 for 3 years; household will pay 10 million VND to the club
3	"Lo O" model	1500 trees (3 ha); this model was carried out in 2009 with the support of agricultural extension station of district. Model was assigned for one household. Benefit sharing mechanism: 70% will belong to the household who manages the model, 30% is given for the club
4	Acacia forest model at grassland areas	3 ha; this model was carried out by investment of members of the club in 2009. Each household contributes 200,000 VND for breed buying and plants; model was assigned to one member of the club for 5–6 years; this household has to contribute 20 million VND (5 million VND paid as soon as receiving model, 15 million VND will be paid after harvesting products)
5	Porcupine raising model	Two households; this model was carried out in 2009 with support of CORENARM through agricultural extension station of district; this project supported 60% (6 million VND) for breed buying, household contributed 40% (4 million VND); household spent by themselves to build breeding cage; this model was carried out in Oct. 2009
6	Indigenous forest trees model	This model has been carried out under the shade of poor forests; it was supported by forestry faculty in 2010

the next models, encourage the relatives of members in the club if they get good results in studying, support a fund to develop production activities for members' households (lending without interest), and visit members and their family when they get ill.

During the process of implementation activities from the club's founding date in 2004 until now, besides the promotion of internal force, the community has received consultation and support of many agencies and related organizations. Consultation and support activities of agricultural extension stations of the district, the SEF, Nature Care, and CORENARM concentrated on three aspects: policies, techniques, and finance. Among them, ability improvement support and internal force rising of the community is the key ingredient that was applied by organizations when activities were carried out. This helped the club change from unconnected activities and results that are not high the first time to become a club with intensive activities and high organization. Forests were protected and well developed, forest development and livelihood improvement models were properly shown, and creating income ability and following the sustainable tendency was done.

10.3.3.3 Generalizing the Factors That Affected the Success of Two Models

Through analyzing two effective forest management models, the process of changing community awareness, handing over the management and forest resources using authority, experiences of the community in establishment and implementation of an internal governance system, and effective support of related stakeholders have been shown. Each model has specific features due to different characteristics of resources and uses by user group. However, positive factors that caused the success of two models are relatively the same, as shown in Table 10.5. The factors that related to the establishment and operation of internal

TABLE 10.5 Generalizing the Basic Factors Affected to the Success of Two Forest Management Models

| | | Model | |
| | | Thuy Yen Thuong | Phu Mau Hamlet's Club |
No.	Factors to Decide the Success		
1.1	Forest allocation policies for community only have been considered if most of members in community have right awareness and understand well the importance of forest management, protection as well as the general value of forest resources	●	●
1.2	Handing over the legal right of forest management when community had clear change from awareness to action, especially they should be assessed and considered by related authorities	●	●
2.1	Establishing the organizational structure follows the purpose of integrating the role of authorities and unions in order to strengthen and increase management power	●	●
2.2	Establishing and modifying the management and protection convention to increase suitably, clearly, and simply implementation of activities	●	●
2.3	Promoting the propaganda to improve the awareness through community meetings	●	●
2.4	Promoting the propaganda to improve the awareness through community activities, culture, and family	●	
2.5	Educating and overseeing effectively the objects who violated in community	●	
2.6	The initiative in understanding and implementing the legal right of forest management board is also an important factor	●	●
2.7	Sharing the benefits fairly, tending to low position people and common welfare activities of community	●	●
3.1	Mobilizing the participation tending to increase the initiative, responsibility, and clear benefits in forest resources management and using of community's members		●
3.2	Clear participation criteria; participants have to show high voluntary and was tied relatively when participated in forest management and protection activities	●	●
4.1	Effective cooperation and support of local forestry agencies	●	
4.2	Consultant and support in policy, technique, finance of related stakeholders		●
4.3	Supporting to establish and develop the livelihood models tending to medium and short terms		●
4.4	Supporting and promoting of related agencies to establish the protection and development fund through the wood harvesting or allow using legally resources	●	●

Note: Above contents related to these categories: (1) Legal framework and tenure rights, (2) building governance system, (3) participation mechanism, (4) necessary support.

governance systems, support and consultation on policy, and technique and finance of authorities play an important role beyond the raising of internal force of the community.

10.4 LESSONS LEARNED AND RECOMMENDATIONS

Based on a legal framework and a common policy of CFM, but due to specifics of test models, different applications for different objects of consultants were created for the diversity in a picture of strategy established and planning to manage community forest in Vietnam, in general, and Thua Thien Hue Province, in particular. Executing the process and obtaining results as well as advantages and disadvantages of practicing were very diverse and abundant in terms of awareness, organizational system, management and protection of forests, community livelihood impacts, and so forth. Many factors in terms of promoting and limiting success of the model of CFM have been detected. Although it is very difficult to confirm which factor is the most important for each case, some factors will stand out *when all of the cases are collected to examine.* Those factors focus mainly on these aspects: legal framework and tenure rights; building institution of internal governance system; getting knowledge about resources and cost of community; and necessary support from related agencies.

The research revealed that to develop the effect and sustainability of CFM models in the study area, the practical issues being considered for implementation are presented as the following points.

First, the beneficial policies for communities that receive forest should be built toward considering the profitability of each state of forest resources and investment of the community. In particular, the issues should include building simple and detailed information on natural resources, adjusting exploitation conditions to shorten beneficial time for the community, estimating expenditures of the community, supporting finance or other beneficial material forms for the community in case of necessity (poor forests, forests that have low-quality wood, poor economic communities). Particularly, policies on advanced timber harvesting should be considered and applied in a timely manner for communities whose forests have satisfied reserves and safeguarded quality.

Second, consolidating and creating a stable legal corridor for the community in the process of implementation of activities as well as how to help them understand their tenure rights, which should be placed in a priority position in the coordination process and support of related agencies. The related agencies should promulgate and transfer decrees and legal documents to the community at the time when they receive forests to manage and use. Documents should be edited easily and transformed to each member in the community. At the same time, activities such as seminars and workshops on policies should be organized to strengthen knowledge and the right synchronous application for the community.

Third, establishing internal governance systems should tend to encourage the participation of the community's members, increase responsibilities of members, and be clear about implementation and monitoring. It includes organizational structure establishment, selection of participated objects, and establishment of operational mechanisms. Organizational structure establishment and selection of participated objects should tend to be on a small scale, connecting power and union roles. Operational mechanisms should be modified in executing the process to consolidate and strengthen the application opportunity. In addition, providing governance ability and conflict management to the community should be carried out before and after forest allocation.

Fourth, the community should consider propaganda to improve the awareness of forest management and protection for members; it is an important activity in addition to forest patrol and protection. Propaganda should be carried out through community meetings (organized by the community or combined with authorities) and other ways such the community activities, culture, and family. In particular, educating and observing the violated objects in the community should be carried out seriously as the regulations agreed.

Fifth, the information on characteristics and the entire value of forest resources should be supplied concretely and in detail for the community. This activity should not stop at supplying the characteristics and amount of wood in the allocated forests but should also include the tendency for the community to identify and develop other values of forest resources if the community knows exploitation and use equipment of documents, and has knowledge about forest caring, nourishing, and developing techniques.

Sixth, communities should build short- and medium-term livelihood models based on the forest resources. This activity will be difficult to carry out without the consultation and support of related stakeholders at the beginning of forest management activities. The support of related stakeholders in identification of the model, technical training, and financial support is really necessary to encourage and promote the internal force of the community. Especially, to promote these livelihood models effectively and sustainably, communities need to build the management mechanisms and benefit sharing mechanisms tending to the assignment for specific members as well as building contribution mechanisms and use of funds.

To carry out these things, the support, consultation, and cooperation of related agencies in terms of policy aspects, techniques, and finance following a comanagement form are really necessary, besides the confidential, proactive, and ability aspects of the community. The support should be carried out at the beginning period when the community is assigned forests and should be regarded as the duty and work of related agencies; the support should tend to improve the ability and promote the internal force of the community. CFM models will not be successful if the paradigm of completely transferring responsibilities in forest management and protection from government to local communities still exists.

References

ALFPU (A Luoi Forest Protection Unit), 2008. Report on Assessment of Forest Allocation Program Within the District.

Bao, H., 2006. Community forest management (CFM) in Viet Nam: sustainable forest management and benefit sharing (in a cut for the poor). In: Proceedings of the International Conference on Managing Forests for Poverty Reduction: Capturing Opportunities in Forest Harvesting and Wood Processing for the Benefit of the Poor, Ho Chi Minh City, Oct. 3–6, 2006.

Donald, A.M., Mol, P.W., Wiersum, K.F., Shepherd, G., Rodriguez, S., Vargas, A., Dedina, S., Stanfield, D., 1993. Common Forest Resource Management: Annotated Bibliography of Asia, Africa, and Latin America. FAO, Rome.

Le, T.H., Ho, T.T., Nguyen, P.N., Le, V.A., Le, T.T.H., Nguyen, T.X.H., Nguyen, T.T., 2010. Community forest management in A Luoi District, Thua Thien Hue Province: current situation and right in practice. In: Common Property Regimes for Sustainable Livelihoods and Poverty Reduction in Central Vietnam, Thua Hoa Publishing House, Thua Thien Hue.

LTCPC (Loc Thuy Commune People's Committee), 2000. The profile of Natural Forest Allocation to Community of Thuy Yen Thuong Village, Loc Thuy Commune, Phu Loc District, Thua Thien Hue Province.

Mahanty, S., Nurse, M., Rosander, M., Greenwood, C., Halley, M., Vickers, B., 2007. Benefit Sharing in the Mekong Region—Lessons and Emerging Areas for Action (in a Fair Share? Experiences in Benefit Sharing From Community-Managed Resources in Asia). RECOFTC, WWF and SNV, Bangkok, Thailand.

NDDARD (Nam Dong Department of Agriculture and Rural Development), 2008. Report on Land Use Planning and Forest Land Allocation.

NDDPC (Nam Dong District People's Committee), 2009. Proposal on Natural Forest Allocation from 2009 to 2015.

Ngo, T.D., Le, T.H., Le, T.H.N., Le, T.T.H., 2009. Institutions and performance of community forest management: multi-criteria analysis framework in a case of forest management in A Luoi District, Thua Thien Hue Province. In: Market, Natural Resource Management and Animal Disease in Vietnam Uplands. Thuan Hoa Publishing House, Thua Thien Hue.

Ngo, T.D., Sakai, T., Moriya, K., Mizuno, K., 2013. Participation in and benefits of community forest management: learning from cases in Thua Thien Hue Province, Vietnam. Geogr. Rev. Jpn 85 (1), 39–55.

Nguyen, Q.T., 2005. Trends in forest ownership, forest resources tenure and institutional arrangements: are they contributing to better forest management and poverty reduction? Case study in Viet Nam. In: Understanding Forest Tenure in South and Southeast Asia. Food and Agriculture Organization of the United Nations, Rome.

Nguyen, B.N., 2009a. Community forest management in Viet Nam. Actual situations, problems and solutions. In: Proceedings of National Conference on Community Forest Management, Ha Noi, Jun. 5, 2009.

Nguyen, Q.H.A., 2009b. Forest Resources Management Through Community Forest Management Model in Thua Thien Hue Province.

Nguyen, T., 2010. Results of evaluation on participatory forest allocation to communities within 10 years in Thua Thien Hue Province. In: Proceedings of Conference on People Right-Based Natural Forest Management, Thua Thien Hue, Aug. 20–21, 2010.

Nguyen, B.N., Nguyen, H.Q., Ernst, k., 2005. Vietnam community forest 2005. First regional community forest forum. In: Proceedings of Regional Forum, Bangkok, Thailand, Aug. 24–25, 2005.

Nguyen, Q.T., Nguyen, B.N., Tran, N.T., William, S., Yurdi, Y., 2008. Forest Tenure Reform in Viet Nam. RECOFTC & RRI, Bangkok, Thailand.

Pham, N.D., Duong, C., 2010. Building and developing the community organization near by the natural forests—effective solution to promote right and duty of villagers in forest management, protection and development. In: Proceedings of Conference on People Right-Based Natural Forest Management, Thua Thien Hue, Aug. 20–21, 2010.

Schlager, Ostrom, 1992. Property-rights regimes and natural resources: a conceptual analysis. Land Econ. 68 (3), 249–262.

TTHDARD (Thua Thien Hue Department of Agriculture and Rural development), 2009. Abstract Report on Forest Allocation Proposal of Thua Thien Hue Province in Period 2010–2014.

VNFPD (Viet Nam Forest Protection Department), 2008. Data of Annual Change of Forest. Retrieved from: http://www.kiemlam.org.vn/Desktop.aspx/News/So-lieu-dien-bien-rung-hang-nam/Nam_2008/.

Vo, D.T., 2010. Benefit sharing mechanism in community forest management in Vietnam. In: Proceedings of Conference on People Right-Based Natural Forest Management, Thua Thien Hue, Aug. 20–21, 2010.

Devising a Guideline for Conservation Education for Secondary Students

M.Q. Huy*, S. Sharma†

*Hue Forest Protection Department, Hue, Vietnam †WWF Nepal, Kathmandu, Nepal

11.1 INTRODUCTION

The wakefulness of the accumulating destruction of the ecosystem's function along with anthropogenic pressure is a growing societal concern. New wisdom is required that can effectively disseminate noble knowledge and both traditional and tacit understandings as new demands in sustainability issues are evoked, thereby generating opportunities for researchers and students throughout the world. Conservation education has been progressing for decades now with the understanding that children are the leaders for tomorrow and they need to be set for tomorrow's challenges and the future they will accede to. This requires a pledge to acknowledge children with environmental and nature conservation education. This conservation education, when incorporated into the school curriculum, will help children to study the environment, expand the habit of exploring issues, and look for the solution themselves. They will also develop skills and attitudes to conserve the environment while they also "understand the principles of ecologically sustainable development."

The overall goal of conservation education is to develop an environmentally knowledgeable community. These environmentally cultured individuals can understand challenges and issues in conservation and analyze how the conservation environment is affected by human decision dynamics. This understanding aids in forming informed decisions and good choices that take socioeconomic and political settings into account. Conservation education builds awareness through formal and informal education on the environment and natural resources to acquire understanding, knowledge, skills, and values to facilitate participation in developing an ecologically sustainable and socially just society.

Throughout the world many schools have carried environmentally friendly activities through directly involving students. Explicitly, in areas around Halimun-Salak National Park in Indonesia a series of conservation education along with environmental-based projects were carried out with active participation from students, headmasters, and teachers. Three years after the project wrapped up, it was found that students were able to undertake similar activities at their homes, which in fact was prominent through many conservation events around the national park at the commune level. Caro et al. (2003) conducted a study among undergraduates enrolled in a conservation course through matching their opinions both before and after the course and found that conservation education was sympathetic to advocating wildlife conservation.

Thua Thien Hue is the first province in Vietnam to consider conservation education as a milestone to promote youth and students toward the management of biodiversity, natural forests, and wetlands. Accordingly, projects were devised to promote environmental education and promote awareness campaigns through establishing pilot models at diverse sites. Multisectoral participation was included through different organizations at the community and provincial level including nongovernmental organizations (NGOs), governmental officials, and educational authorities. Basically the Forest Protection Department and Nature Care Association (an NGO) had a major role to play on raising awareness.

This chapter considers one of the earlier forms of conservation education courses, "Nature Conservation Education Program," as an initiating point. The course was conducted for 4 months in Aug. 2007 and involved a secondary school and four villages of Phong My Commune in Thua Thien Hue Province. The overall goal of this course was to study conservation education concepts and processes from a multidisciplinary perspective to stimulate individual student's learning on a conceptual learning basis. This chapter discusses the course along with individual experiences of involved participants. This is then discussed in terms of how interpretations and individual experiences from the course narrate how academic concepts may be valuable in further evolving "conservation education" courses. The key emphases is on highlighting the steps needed to enable more productive development and utilization of conservation education in terms of concepts, approaches, and tools.

11.2 METHODOLOGY

The overall objective of this chapter is to gather, analyze, and understand feedback from related stakeholders to form a guideline that is likely to involve community educators in both formal and informal education among the Phong My secondary school and four villages of Phong My Commune: Tan My, Luu Hien Hoa, Ha Long, and Khe Tran. Though there already exists a similar course focused on conservation education, this chapter merely discusses methodological improvements.

Both qualitative and quantitative studies are considered appropriate as the focus is on stakeholders' understanding of particular context and further discovery of situation insights. Stakeholders' feedback from school leaders, teachers, students, the education office, parents, the youth union, the children's council, and the head of communes were taken through diverse interviews with the objective to gather representative opinions on developing conservation education programs for suitable audiences that provide benefits to the community. Guiding stakeholders' interviews was backstopped by individual surveys, determined by using a formula suggested by Yamane (1967). Sample students and teachers were calculated at the 90% confidence level with precision $e = \pm 10$. The total sampled teachers and students were 21 and 82, respectively. Activities undertaken under the current nature conservation education program were individually analyzed to develop a guideline that represented local context and expertise and formed guiding material to mainstream a conservation education program efficiently in the future.

11.3 THE EXISTING COURSE

The earlier course, "Nature Conservation Education Program" for secondary school, was designed to provide an opportunity to develop the conceptual framework of conservation education to enable students to be more constructive, integrative, and innovative in their work, including diverse socioeconomic and environmental factors. The overall objective was to render help to students and teachers to frame a wider overview of the tools and concepts of conservation education that happen in school. Because the course needs to be further improvised through a new developed guideline, the existing practices are carefully understood.

The practical course was organized earlier through green clubs. Green clubs are preferably run by school volunteers who undertake practical environment-related activities targeted for secondary school students. Green clubs are believed to contribute to a comprehensive education program to consolidate students' knowledge and skills in an attractive way usually carried out during after school hours. Experiences from the past show that green clubs form a playground to help students develop solidarity and group working skills. Usually schools in rural and periurban areas least prioritize extracurricular activities into educational systems. As such, establishing these clubs provides access to practical learning in science and environment around them. The students involved in green clubs also belong to a society where the message on conservation could be disseminated while adults concomitantly get involved in the process too. In a way, formation of green clubs targets both the school and the community to become more responsible and act proactively in matters related to environmental protection and natural resource conservation.

11.3.1 Results From Stakeholders' Interviews

11.3.1.1 Support for Conservation Education Programs

There was a positive response from multiple stakeholders in developing a conservation education program for diverse reasons. The first and most prominent reason was that these courses will provide rare opportunities for practical environmental learning. Additionally, knowledge obtained through these learning modules needs to be shared with the public that does not have access to professional articles and journal publications. Some interviewees also said that giving meaningful outdoor experience would create wider understanding of the existing world and aid the younger generation to care for the surrounding environment and act to lower the risk accordingly. Moreover, creation of conservation education fits into Vietnam's long-term goal to advance research and learning.

Though local educators and students showed solidly positive replies on the need for conservation education programs, they highlighted the need for technological improvement in units for undertaking such conservation learning modules.

11.3.1.2 *Target Audience*

Wider responses on who should be the target audience for conservation education were observed. Many respondents believed the program should reach as many individuals as possible, while others identified secondary students, teachers, youth clubs, and community mobilizers to be the primary target for the course.

The geographic location was felt to be equally important together with targeted youth as the primary audience. Some respondents identified city youth and students as they are least exposed to the external natural world and for their capacity to advocate the necessity for conservation. Major respondents emphasized students and youth in the surrounding buffer and conservation zone as the primary audience because they are closer to the incidences of poaching and illegal harvesting that are major problems leading to unsuccessful conservation. Similarly, teachers and educators were mentioned as a distinct target audience both at the local and national scale, with the understanding that educating teachers is likely to disseminate learning more easily. Equally important are university researchers and students, donors, social clubs, and business groups. It will be an opportunity for these groups to conduct research in conservation areas to develop ways to intervene in providing ideas for overall management of conservation areas.

11.3.1.3 *Proposed Program Goals*

The respondents listed the following goals on which a future conservation education program should focus:

(a) Creating awareness and understanding of the existing environment.
(b) Creating positive attributes toward the environment for motivating individuals to act and improve their behavior to protect it.
(c) Developing skills for identifying and addressing conservation issues.

Starting with creating awareness, learners should begin with concrete steps until environmental literacy and willingness to act to protect environment is achieved. Local instructors saw a dire need for such an education system in their local community. A common statement from the respondents was that local students inhabiting conservation areas or buffer zones must have more conservation education programs. This is because "these are one of the dynamic areas and students don't seem to care much," as explained by one of the respondents.

11.3.2 The Gap in an Earlier Method of "Nature Conservation Education"

The actual application of conservation education is rather limited. To foresee substantial environmental changes, core academic transformations alone are not sufficient to reduce the lack of environmental and conservation compliance. A transformative environmental conservation learning process is required within the children's education system from its early stages. Issues like the lack of human resources that can advocate for the importance of conservation education are persistent. Human resources with the dual capacity to acknowledge children's psychology and conservation experience are scarce. Simultaneously missing is the institutional arrangement through which these interventions could be implemented. At a higher level, the lack of a proper mechanism to integrate conservation education into the formal school curriculum is a major issue. This is further aggravated by reduced availability of educational reference materials written in simple and easily understandable languages. There are enough educational materials written at a national level and not necessarily applicable locally, owing to complex content and language. Environmental education is based on values and perception. Values and perception vary among individuals and pose difficulty in initiating with a singular ideology often prominent through lack of efficient public participation and networking in environmental initiatives. Even at schools, hard-core academics is prioritized while environmental conservation education is least considered. This needs to be expanded more from biology and ecology to more practical-oriented educational modules. Today, a number of tools for conservation education exist but they are not adequately implemented. Institutional cooperation and coordination in planning and implementing such projects is required to establish a network between students and nature.

Nature conservation education is essentially required in the buffer zone communes of newly established national parks and conservation areas in provincial zones to create youth awareness and mobilization. However, earlier experience showed that activities undertaken were not systematic in Vietnam and there needs to be a guiding document that is contextually relevant and appropriate where learning and topics are considered. This guiding document should consider values and perceptions that aid in mobilizing and engaging through better participation. In this regard, this chapter considers the process of guideline drafting for conservation education focused on secondary school children and teachers to equip them with comprehensive tools to promote a practical conservation environment. If proved successful, it can be integrated into the forest policy as one measure to conserve forests. Attempts have been made to capture complexity responsiveness among varying social, economic, and environmental contexts within which youth and schoolchildren live. Endeavors to improve open learning and teaching programs are incorporated in such a way that children and youth are motivated through developing specific skills, practices, and strategies toward environmental and conservation issues.

11.4 PROPOSED GUIDELINE FOR FIRSTHAND CONSERVATION EDUCATION PROGRAMS

A guideline was prepared based on the information gained and gaps acknowledged in earlier programs. This "guideline for conservation education" provides educators, students, and administrators with a voluntary guideline to enable conservation education programs within their school. This guideline will backstop school and local conservation efforts by

(a) Setting prospects for performance and accomplishments through strengthening green clubs at individual schools.
(b) Creating an outline for operative and comprehensive conservation education courses and prospectuses.
(c) Signifying how conservation education meets criteria set through customary disciplines that provide students with ways to accumulate knowledge and synthesize experiences among diverse disciplines.
(d) Outlining crucial goals for conservation education.

The proposed guideline is believed to set quality conservation education diagonally through conservation and buffer zones, based on a standard of what a conservation-literate individual needs to be doing.

11.4.1 Vision

"Guidelines for conservation education" is grounded in shared an understanding of conservation education that agrees with the major founding views of the field (as mentioned earlier) and follows international supervisory documents through the Belgrade Charter and the Tbilisi Declaration (NAAEE, 2000). A United Nations conference adopted the Belgrade Charter through providing a recognized goal/aim statement for conservation/environmental education:

> The goal of environmental education is to develop a world population that is aware of, and concerned about, the environment and its associated problems, and which has the knowledge, skills, attitudes, motivations, and commitment to work individually and collectively toward solutions of current problems and the prevention of new ones. *(NAAEE, 2000)*

The Tbilisi Declaration was adopted a few years later and established a wider aim for conservation education, which later provided a baseline in similar ground since 1978:

(a) To raise awareness of socioeconomic and ecological interdependence among rural and urban areas.
(b) To foresee providing each individual with opportunities to gain the values, attitudes, knowledge, and commitment required to improve and conserve the surrounding environment.
(c) To promote new behavioral interaction patterns among individuals and groups as a single entity among the overall environment.

With this proposed guideline, the abovementioned ideologies have been critiqued and revisited. However, they still stand for the foundation for shared knowledge and core concepts of environmentally literate individuals. Conservation education is rooted in the concept that humans and nature can act equitably, while decisions made by humans can be informed to consider this linkage for future generations.

11.4.2 Establishing Green Clubs in Schools

Green clubs are preferably run by school volunteers who undertake practical conservation-related activities targeted for secondary school students. Green clubs are believed to contribute as a comprehensive education program to consolidate students' knowledge and skills in an attractive way, usually carried out after school hours. Experiences from the past show that green clubs form a playground to help students develop solidarity emotion and group working skills. Usually, schools in rural and periurban areas least prioritize extracurricular activities into educational systems. As such, establishing these clubs provides students with access to practical learning in science and the environment around them. The students involved in green clubs also belong to a society where the message on conservation can be disseminated while adults concomitantly get involved into the process too.

In a way, the formation of green clubs targets both the school and the community to become more responsible and act proactively in matters related to environmental protection and natural resource conservation.

11.4.2.1 Structure of Green Clubs

Usually, depending upon the number of students and teachers in a particular school, one or two green clubs can be established within each secondary school. The process of forming green clubs needs to be supported by educational staffs from nearby conservation areas and buffer zones. An individual green club may have a maximum of 40 members and 2 teachers to manage them. Implicitly, with more than 40 students the groups could be divided into 2 groups with 20–40 members each, as a higher number of students would pose difficulty for managing students. This is the optimum number that allows students to work and play together with equal benefit and participation. Nevertheless, the greater the number of students participating in green clubs, the more likely the conservation program is to be more effective. However, attempts should be made to maintain a 1:20 teacher-to-student ratio. In conditions with 60 participating students in green clubs, according to the ratio of one teacher for every 20 students, following the same curricula and activity is possible. During field trips these groups may travel together but require supervision by individual teachers.

At secondary schools, green clubs work best for students from grade six to grade eight, because grade nine students are usually under the pressure of high school entrance preparation. Also, focus should be given to local students as they are familiar with local customs and instances. Regular meetings at regular intervals help in better planning; however, the time and length of each meeting should not conflict with regular school hours.

11.4.2.2 Members of the Green Club

All the green club members are volunteers acting importantly as a leader, secretary, photographer, artist, singer, reporter, researcher, and so on, sharing knowledge, information, and experiences with each other. The position of head and secretary is taken by two students, respectively. The head manages the club and invites all the members to meetings. He or she also has the responsibility of supporting schoolteachers in organizing conservation education activities. The head also maintains and consolidates solidary spirit and unit among themselves while also proposing vital suggestions to teachers. The secretary helps in composing a newsletter, keeping meeting minutes, and supporting the head in formulating learning tools within the club.

11.4.3 The Facilitator

Activities within conservation education are organized in diverse forms mostly out of the classroom. Facilitators need to have knowledge and practical experience, so that student-lively and attractive conservation activities can be organized in school. A facilitator plays an important role not only as an educator but also acts as a director (stage director), editor (scenario writer), game controller, storyteller, master of ceromony (MC), organizer, performer, propagandist, defender, and player to make activities lively among students. Within the nature conservation course many capacity-building and awareness-raising activities should be conducted in the form of special dramas in which green club members play major roles while the facilitator manages the overall process. Conservation education activities involve special dramas, in which facilitators prepare scenarios and plan them so that the facilitator can use it to implement activities accordingly. Green club members often learn by participating in different learning games. It is the facilitator's role to explain the rule and maintain a lively atmosphere within the process. A good facilitator should be able to coordinate activities suitably and provide crucial comment at the correct time to render wider attention and interest among students and surrounding beneficiaries.

Facilitators are basically educators, and the motto for such conservation education programs should be "Learning by playing, playing by learning" with the basic criterion of "fun studying to conserve nature." They are required to design activities in accordance with children's psychophysiology and situation reality to which knowledge on principles and methods on environment and experience in dealing with children are musts. In the process, students may discover sudden events and queries; and it is the facilitator's role to make it easy for them to understanding the inquiries.

11.4.3.1 Criterion for Selecting Facilitators

Facilitators are basically the teachers managing the green clubs; as such they are required to be well acquainted with the environment and child physiology. They are also required to be well informed about the local surrounding environment and conservation issues. They must be willing to take challenges, show innovative pathways to children, be creative as problem solvers, and most importantly, be role models for students. Each facilitator is required to respect young ideas and look for ways to nurture them further. To ensure a good facilitator for promoting conservation education, the following criteria could be applied:

(a) A person willing to voluntarily participate in conservation education.
(b) Individual experienced in conservation issues and child psychology.
(c) Socially active and bearing the capability to convince a group of people.
(d) Individual trusted by the school and the surrounding community.
(e) Must be a role model with a good attitude toward one's duty and surrounding environment.

11.4.4 Organizing and Implementation

Defining the particular problem is the first step in designing a course. Prior to defining the activity to be conducted under the course, a problem tree is first designed to identify causes and effects for any specific issues, with the understanding that the more proximate the problem source being addressed, the more successful the approaches will be. The problem is then analyzed and translated into specific activities to target as a conservation education program. The activities can be conducted anytime within the classes, surrounding schoolyards, and as outdoor events depending on the number of students and weather. Children often find outside activities to be more fun so that much learning could be extracted and they can participate in outreach activities in their respective village. In particular, there are many reasons to focus on nature as important in the classroom for conservation education, as follows:

(a) Curiosity within students arises more in nature than in the classroom. Outdoor learning encourages students' inquisitiveness.

(b) To stimulate added interest in a given area or unit of study.

(c) Through outdoor activities, students may collect specimens of flowers, leaves, fruits, roots of plants, insects, soils, rocks, and so on that may be brought back to their classroom for additional study.

(d) Students who observe and record data on field trips mostly will be able to propose reasoning for differences in animal and plant growth at different stages, environmental impacts of industrial flows, and much more.

(e) Students may visit and study historic relics to gain an appreciation of historic events.

(f) Students will be able to utilize their free time effectively through developing the capacity to differentiate man-made and natural free-time opportunities.

(g) Students understand our world in small bits and pieces so that integral parts of the total picture are broadened.

11.4.4.1 Preparation by the Teacher

Teachers or facilitators must be active enough with strong planning before taking students out for outdoor activities. In particular, facilitators need to make a quick visit to the field site before actually taking a trip with students, to avoid potential problems that may occur. Simultaneously, parents and the head of the school need to be acquainted with the details of the trip. Local guidelines need to be explained beforehand so that the explanation is appropriate to a student's age and the standard of the question likely to be asked, as well as the available time and materials needed. Also, important materials for the trip are a special toolbox, protective student clothing, and sufficient student food and drinks. A strong traveling plan is comprised of safety measures together with better strategies for observation, information recording, and maintaining a time log. Depending upon the number of students taking part in outdoor visits, a concomitant number of teachers should be maintained accordingly. An attempt should be made to maintain an 8–10 student per teacher ratio to help assist students during group work that is mostly conducted outdoors.

Students are keenly interested in outdoor activities, mostly looking for events outside schools. Once they are out, the tendency exists to become overexcited, so teachers need to be extra careful while handling them. At the same time, students need to have a basic understanding of a problem at the one hand while they correlate the learning with the school curriculum and problems in the surrounding environment on the other hand. Students need to be involved in such a way that they make careful observation, make exact recordings, make critical reviews, and evaluate accordingly. Before taking part in an actual field visit, students should practice at proximal sites on how information should be gathered and things to be studied while in the field, as these are skills that require a long time to learn. Students must know the objective for which the information is being collected and it is the teacher's role to explain that outdoor activity provides students with opportunities to apply classroom learning. Students must feel that an outdoor activity will be interesting and there will be much to learn; not just another lesson, but a set of facts. For each and every student to participate in the learning process, a specific responsibility should be given to each student and it is the facilitator's role to facilitate the overall doings and respond to every student's expectations. Informed decision making requires that common ground rules be set up through discussions.

Individual people will have different field experiences and diverse objectives for learning. It is the facilitator's role to support individual learning approaches through being more enthusiastic and guiding students. Each student has a different capacity to convert observation to learning, so enough time should be allocated. The capacity for sustained interest varies, and the pace must take interest spans into consideration. Better planning always facilitates process toward better culture, so attempts should be made to orient new surroundings in a way that is comfortable and simultaneously improve students' attitudes and behavioral patterns.

11.4.4.2 Evaluation and Follow-Up

Conservation education is fundamentally dealing with child psychology. If the facilitator shows no interest in activities undertaken by respective students and fails to encourage them to express their reactions, it is a failed program. Timely follow-up helps both facilitator and students; the specific vision required is accomplished time and again. After field visits, students should be encouraged to seek answers to issues they discovered. Writing exercises need to be promoted so that students can better describe events and activities. Writing can embrace drawing and/or forms of expression that could be shared with facilitators and students' own parents. It is wise to recognize that the social experiences of being with a group other than the family may be enjoyable and offers another form of understanding and learning that could be further related to classroom activities individually through reports, projects, demonstrations, and presentations. Prior to doing this, associated queries experienced by individual students during field work needs to be discussed with the students by the facilitators. Whatever the purpose, outdoor activities should always be preceded and followed by both thinking and planning.

11.4.5 Designing Individual Models Under Conservation Education for Students

Each module is an independent unit within the conservation education subject developed to provide a guideline for implementing the overall conservation education subject. The module is developed in a flexible way so that one module could be developed for each activity, which are generally practical field experiences taught beyond classrooms. Facilitators may use particular modules for subjects beyond conservation education to aid bigger events.

Designing and organizing a conservation education module guideline is a difficult task to incorporate basic demands from the stakeholders that match students' psychology and can be easily achieved. It is often regarded as beneficial to design modules in combination with associated theories and information coupled with discovery games.

To ensure a good conservation education module, the principles of 3A (Awareness, Attitude, and Action), 3Co (Cooperation, Coordination, and Communication), and 3R (Reduce, Reuse, and Recycle) need to be incorporated, which reflect the real situation of the school, club, and locality that matches the student's ability to pursue and understand. Attempts should be made to make the module flexible, attractive, and joyful.

11.4.5.1 Components to be Included Within an Individual Module

The guideline for devising modules under conservation education for students is divided into four sections, where individual sections contain questions envisioned to assist the facilitator in designing a particular module and statement for its conclusions. This guideline may be valuable for mid-career professionals and researchers. With a view to avoid gender bias, the guideline is represented in the second person and also includes views from stakeholders' interviews conducted earlier.

SECTION 1: GENERAL INTRODUCTION

1. Has the guideline adequately described associated problems and theoretical orientations?

    Every student is influenced by values, family background, and theoretical orientation. During stakeholders' interviews, participants stated that students acknowledge their personal characteristics and reveal them decently while undertaking any conservation-related activities. In this regard, a consumer of green clubs incorporating conservation-related activities should be assured that understanding more beyond the scholarly fashion in a way students will be guided well in terms of conceptualization and execution.

2. Has the learning been extracted from wider literature that is significantly representative?

    Stakeholders' interviews revealed that facilitators involved in executing conservation education activities should be able to demonstrate considerable familiarity with similar activities conducted elsewhere.

    According to WWF-Nepal (2015), there are more than 500 ecoclubs that conduct activities similar to green clubs, with more than a hundred thousand children participating with almost half of the members being girls. They conduct diverse activities including trees plantings, organizing campaigns for environment cleaning, and most importantly raising awareness on conservation in the form of essays, paintings, and speech competitions.

SECTION 2: GENERAL IMPLEMENTING PROCEDURE

3. Is the choice of a particular activity appropriate?

    Stakeholders' interviews revealed that the facilitator should present a rationale for conducting a particular activity regarding potentiality to existing issues and challenges that are relevant in the locality.

4. Was there an adequate wealth of data collected?

    Stakeholders felt that enough data should be collected to maximize opportunities to answer prevailing issues and challenges. Recording suitable information could be made easier through developing appropriate forms that match individual subjects compatible with the student's ability and psychology. It also provides a comparable baseline for conducting similar activity the next time.

    After undertaking any activity, it becomes crucial to record the information in a way that can be used in the future for diverse purposes. The stakeholders highlighted the activity forms mentioned in Box 11.1 as a prerequisite for data collection purposes.

5. Have the students been adequately informed about the need for a specific activity?

    Students must be very familiar with why a particular activity is being conducted to ensure that they will be interested and are able to feel the importance of the activity accomplished. Given the complexity and inconsistency of an individual student's character, the same activity may not have equal level of importance for a given student. Stakeholders' interviews suggested that given a full explanation be given to the students, so they will be able to assess the related importance themselves.

BOX 11.1

SOME EXAMPLES OF ACTIVITY FORMS

- Rules for educational games
- Forms for drawing and painting pictures
- Guidelines for making plastic arts with pictorial illustrations
- Forms of making a picture with trees and leaves
- Making toys from waste materials
- Collections (of pictures, environmental songs, good ideas for environmental protection, initiatives, experiences, medicinal herbs, tree leaves shapes, etc.)
- Photo report about environment and nature (presentation with different way)
- Environment survey forms
- Crossword puzzles
- Drama scripts, etc.

SECTION 3: ACTIVITY OUTCOMES

6. Did you focus on why sufficient information was required to permit students' self-interpretation?

    Stakeholders agreed that it is likely to be counterproductive to explain why information and details were required so that students themselves understand the characteristics of the information and will be able to present them correspondingly.

7. Have you clearly described relationships between the activity and outcome(s)?

    There is a temptation by some students to misconnect between activity and outcomes due to misinterpretation. To avoid such misinterpretation, it is the facilitator's role to clearly make them understand the link between the activity conducted and the outcome they might have.

SECTION 4: DISCUSSIONS

8. Has there been a reasonable interpretation of the outcome(s)?

    Students should be made aware that they should not interpret more than what they observed and the information they collected. Students should be able to intend to share this learning and observations among their friends' circle, family, and class. It is the facilitator's role to help the students make a reasonable interpretation of what they have perceived.

    Apart from these major questions, some basic attributes are also required to be considered in the guideline. They are presented in Box 11.2.

    The facilitator needs to evaluate and summarize the overall process of implementing individual modules. This should also include the level of awareness and attitude generated through implementation of a particular module. A group discussion with students before and after the individual activity eases the process. A simple module could be designed if facilitators lack experience in the field. It is not crucial to follow each and every step as mentioned above to create a successful module, but these may serve as a vital guiding factor. It is up to the facilitator to explore a student's activity as a way to seek ideas to set up modules. Whatever may be the content and context to successful module preparation, student safety is vital.

BOX 11.2

BASIC ATTRIBUTES REQUIRED TO BE INCLUDED IN THE GUIDELINE

Objective

- Objective identification, rationale, and how individual modules have been organized.
- Ensuring that opportunity for school students to access knowledge on ecosystem and environment, associated issues, and then looking for ways to address these issues is created.

General introduction

- Context, content, knowledge, and information that will be transferred to students.
- Module should use pictorial illustrations for individual subjects. For example, the diversity in terms of leaves, occurrence of forest fires, etc.

Time and place

- Usually, conservation education must be carried out in between school work so there should not be time conflicts with class hours. As such, better participation could be foreseen.
- The module should clearly identify duration required for individual activity and overall module. Usually, shorter modules are easy to accomplish.
- Some modules can be conducted among community children so the venue for individual models could be arranged in a way to promote multiparticipation.

General Preparation

- Materials, equipment, and associated techniques are required to be efficiently rendered.
- Prehand communication with required technical stakeholders for information support should be prepared correspondingly.
- Illustration materials like pictures, maps, magnetic compass, and measuring tape should be collected in a timely manner.

11.5 CONCLUSION

The guideline proposed herein is incomplete as students are likely to do best with what they are equipped with. The key to using any guideline rests with the need and context itself, as the previously published guideline was not consistent with what was needed, especially at Thua Thien Hue Province, as they have already promoted youth and students toward biodiversity management, but this guideline will only help them to organize better. The interviews conducted among stakeholders indicate that many of the issues could be better managed with a dialog, which is something additional the facilitator should promote.

Designing a guideline for nature conservation education is important in the way it equips necessary preconditions for successful implementation in schools. This can be a good reference material for environmental managers, conservationists, environmental researchers, lecturers, and students of natural resource management. This guideline aids in solving difficulties encountered by educators due to the lack of guiding material. However, this guideline needs to be further improved in a way that nature and human psychology are dynamic and associated issues that vary concomitantly. This guideline serves only as general supervisory material; facilitators/educators must improve other training materials based on local needs and demand. The use of local and regional environmental materials needs to be encouraged. Specifically, educators should initiate new ideas to organize attractive and effective conservation ideas. The implementation of conservation education requires active participation of key stakeholders in the provinces and districts. A network should be established and coordinated with suitable representatives from relevant organizations.

References

Caro, T., Mulder, M.B., Moore, M., 2003. Effects of conservation education on reasons to conserve biological diversity. Biol. Conserv. 114, 143–152.

NAAEE, 2000. Guidelines for the Preparation and Professional Development of Environmental Educators. North American association for Environmental Education. Retrieved from: http://resources.spaces3.com/5e156799-5cd9-406e-835d-748cce277ecf.pdf

Yamane, T. 1967. Statistics: An Introductory Analysis. New York: Harper and Row.

WWF-Nepal. 2015. The Terai Arc Landscape Project (TAL)-Eco Clubs: Learning from an early age. Retrieved from: http://www.wwfnepal.org/about_wwf/conservation_nepal/tal/project/eco_clubs/.

INSTITUTIONS AND POLICY DIMENSIONS

Comanagement Approach for Conflict Management: A Case Study of the Phong Dien Nature Reserve

N.D.A. Tuan*, N.T. Dung†, S. Sharma‡

*Forest Protection Department of Thua Thien Hue, Vietnam †Hue University of Agriculture and Forestry, Hue, Vietnam ‡WWF Nepal, Kathmandu, Nepal

12.1 INTRODUCTION

Until the 1990s, forests were still under state management in Vietnam. With the introduction of the *doi moi* (renovation) process, a decentralization approach brought together multistakeholder participation in forest management. In line with this changing institution, protected areas (special use and protection forests) remain under state management, whereby production forests were allocated to local people through long-term tenure arrangements. While forest quality is higher in protected areas, most allocated forests are degraded and fetched less value to local communities. This is one of the reasons behind local people trying to access nontimber forest product (NTFP) and restricted timbers in protected areas. The Phong Dien Nature Reserve (PDNR), soon after its establishment in 2003, came across a major conflict between local communities and reserves' officials. The officials were granted the right to conserve the reserve's ecosystem and reduce its degradation through prohibiting local people in the core zone. The local communities, having no alternatives for income generation, were dependent on the reserve for their livelihood. This preference difference caused a series of conflicts as local people continued haphazard hunting and logging inside the protected area. Coupled with incidences of forest fires and shifting cultivation, forest inside the reserve was eventually fragmented followed by the extinction of endemic species. In this regard, this study aims to analyze causes of conflict and seeks the possibility of comanagement as a sustainable approach to enhance the good relationship of the park and the people. Findings were expected to help PDNR officials to resolve conflicts with local communities while simultaneously improving their livelihoods, reducing pressure on the natural forest, and developing a conservation strategy for the future.

12.2 RESEARCH DESIGN AND METHODOLOGY

Straddling over two districts of Phong Dien and Aluoi in Central Vietnam, the PDNR hosts 27,000 individuals through 5000 households among 9 communes in the buffer zone. Most of the residents live in the east of the reserve where transportation is easy and the arable land is fertile. The core area covers 41,433 ha and the buffer zone comprises 43,600 ha (Fig. 12.1). The PDNR protects the lower reaches of the Bo, Olau, and My Chanh rivers and provides economic benefits to local communities through its abundant NTFP species.

This protected area falls entirely within the biologically important Central Annamite Range, listed in 200 global hot spots of biodiversity (Tordoff et al., 2003). The PDNR is also known for its wide range of biodiversity and diverse ethnology (one major and five minor ethnic groups). Nevertheless, the reserve is threatened by human activities, which have shown the necessity for priority species and lowland forest landscape conservation, especially in the region between central Truong Son and the border vicinities of Laos and Vietnam. It is one of the few remaining habitats for the endangered Edwards's pheasant (*Lophura edwardsi*) as well as other endemic birds such as *Arborophila merlini*, *Jabouilleia duajoui*, and *Rheinartia ocellata ocellata*. The reserve also hosts endangered large mammals such as saola (*Pseudoryx nghetinhensis*), tigers, and gibbons (Le Trong Trai et al., 2001).

This study applied participatory learning and action (Ashley et al., 2013) methods for analyzing park-people relationships. Specifically, it uses context analysis to help stakeholders

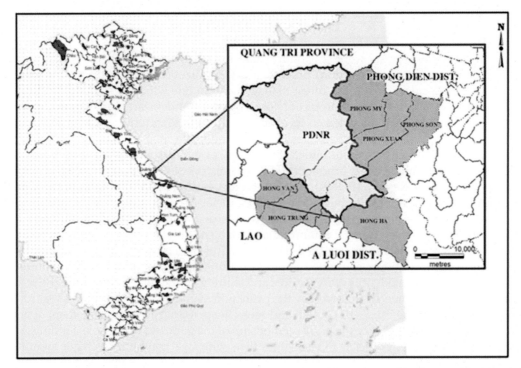

FIG. 12.1 Location of Phong Dien Nature Reserve (core zone in red boundary and buffer zone in pink color) in Central Annamite Range.

examine the origins and underlying causes of conflict and to identify the issues that have contributed to the conflict. The study also applied stakeholder analysis (Grimble, 1998) to identify and assess the dependency and power of diverse stakeholders in a conflict. In addition, the Four "Rs" analysis (Mayers, 2005) was used to examine the rights, responsibilities, returns, and relationships among different stakeholders in relation to natural resources as a means of improving conflict understanding. To visualize resources and associated conflicts, the study used a technique of mapping conflict over resource use to show geographically where land or resource use conflicts exist or may exist in the future.

12.3 RESULTS AND DISCUSSION

12.3.1 Causes of Forest Degradation in the Phong Dien Nature Reserve

Several studies have shown causes of forest degradation in the PDNR (Tran et al., 2002; Le et al., 2001; Tordoff, 2002). Here is the nonexhaustive list of the most direct and indirect causes of forest degradation in the PDNR.

- *Illegal logging*: In PDNR, logging occurred mostly for two purposes: explicitly for trade and local consumption. Trade-driven logging was carried out by both outsiders and local people. Logs were removed along "sled" tracks by buffalo, along streambeds, or flowed in the rivers. For local consumption, people logged timber from the reserve for their house construction and fulfilling other household needs.
- *Hunting and trapping*: Hunting was primarily for illegal commercialized ventures targeting specific key species such as tigers, bears, turtles, and porcupines. People also hunted for fulfilling local diet supplements, income generation, and sometimes for fun. Trapping was more widespread and pervasive, being an indiscriminate and common practice employed by nearly all forest users. Both activities were driven by trade demands for bush meat, medicine, and ornaments.
- *Illegal mineral extraction*: Since early 2003, an influx of gold seekers has resulted in habitat disturbance, increased hunting and trapping, and pollution of streams and waterways within the reserve. The other major activity was scrap metal collection, which is not only dangerous to human lives but also threatens forest integrity as collectors often lighted forest fires.
- *Habitat conversion and agricultural expansion*: These were attributed to agricultural pressures, forest fires, resettlement, in-migration, lack of land-use planning, and lack of clear landownership. Forest is converted to agriculture with minute increment in population. There has also been some access to the construction industry and a new economic zone has been opened on the northeastern boundary of the reserve.

When analyzing the abovementioned causes of forest degradation, we discovered that there were no studies on resources uses before the establishment of the PDNR. More importantly, the feasibility study for the establishment of the PDNR focused mainly on conservation and ignored community development and awareness rising among local communities residing in the buffer zone areas. In other words, neither existing local practice nor potential impacts on local livelihoods were taken into account during reserve establishment. This can be the root cause of conflicts between local people and the reserve as well as with other external stakeholders. These problems were exacerbated by the lack of planning to attract community participation in managing the reserve together with a top-down approach of decision making that marginalized local people's needs and expectations.

12.3.2 Timeline of the Conflicts

People in the PDNR's buffer zone have a long history of living near forests and managing them for their livelihoods. They have accumulated considerable experience and knowledge about local forests and biodiversity conditions.

In 1996, an endangered bird called Edwards's pheasant was rediscovered in the mountains of Phong Dien District after 76 years of disappearance in Vietnam. This bird was first discovered in 1895 in Central Vietnam. This was one of the main reasons for establishing PDNR l as it was the only habitat for the bird.

The grounds for conflict were laid in 2000 when investment plans were initiated for the establishment of the PDNR to conserve Edwards's pheasant and its habitat. The government prohibited logging in the core zone of the reserve. The situation deteriorated in 2003 when the reserve was officially established with its forest ranger team. Conflict peaked in Mar. 2006 when entry into the core zone was strictly forbidden. Forest rangers pinpointed violations including illegal exploitation of timber, destruction of the natural forest, and illegal hunting or trading of wild animals. Violators were prosecuted by the reserve's management board. Local people negatively reacted to the prohibition rules by setting fire to forest within the reserve. The situation has cooled down owing to improved communication, information dissemination, awareness rising among local communities, and community capacity building on sustainable utilization of natural forests (Table 12.1).

TABLE 12.1   Key Milestones of Conflicts Between the Reserve and Local Communities

Time Line	Events	Conflict
Before 2000	A long history of residents living near the forests and using them to sustain local communities In 1996, rediscovery of Edwards's pheasant in the mountains of Phong Dien District	None
2000	Investment plan for the PDNR to conserve Edwards's pheasant and its habitat. The government prohibited logging	Local people were worried about their livelihoods. Conflict started
2003	PDNR established in accordance with Decision No. 2470/QD-UB on Aug. 2, 2003, by Thua Thien Hue Provincial People's Committee	Conflict worsens
2006	Personnel of the Management Board take charge in Mar. Entry to the core zone was strictly forbidden	Conflict peaks
2007–09	Some programs were implemented to mitigate conflict by both internal and external actors	Conflict abates but remains unresolved
2010–13	Implementing buffer zone programs and supporting local communities	Mediation of the conflict begins

12.3.3 Main Stakeholders in the Phong Dien Nature Reserve

There were three main stakeholders in the context of the PDNR: the reserve's management board, local communities, and the Commune People's Committee (Commune PC). In addition, there were other mass organizations such as the women's union, the farmer's union, and the youth's union who also participated in forest use and management in the study area. Roles and relationships among key stakeholders in managing natural forests are presented in Fig. 12.2.

12.3.3.1 The Local Communities Who Are Living in Buffer Zones of the Reserve

Local inhabitants' lives were highly dependent on natural resources through daily activities such as hunting wildlife, and consuming forest products for fuel, medicine, and daily meals. Limited access to arable land or having small areas of garden, local people often encroached in the reserve for harvesting commercial products or sometimes practiced shifting cultivation with cassava and upland rice.

The local community has taken part in varied aspects in preserving and developing natural resources. Because their lives were attached to forest uses, they have accumulated abundant knowledge on species, habitats, and landscapes of the reserve. These characteristics made local people perfect agents to protect, manage, and develop forest resources within

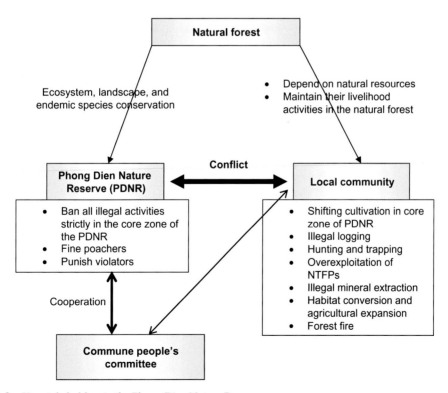

FIG. 12.2 Key stakeholders in the Phong Dien Nature Reserve.

the reserve's boundary. Realizing the vital role of the community in the conservation, the management board of the PDNR carried out many activities to raise people's awareness and support community livelihood in the buffer zone to help improve the resilience of both local communities and natural resources. During the period of 2003–09, a project supported by the MacArthur Foundation was implemented to reduce unsustainable uses of natural resources, introduce appropriate practices for forest management, and harmonize conflicts between development and conservation objectives. Results included effective operation of ecological activities in the buffer zone, building ecovillage models with sustainable development strategies, and developing a social network through clubs of "nature for life" among adults of villages located in the buffer zone of the reserve.

12.3.3.2 *The Phong Dien Nature Reserve's Management Board (The Reserve's MB)*

PDNR officials protected and managed the core zone. They identified violators and enforced various sanctions depending on the level of impacts. As stipulated in Decision No. 2470/2003/QD-UB issued by the People's Committee of Thua Thien Hue Province, the mandates of the Management Board of PDNR included:

- Manage, preserve, build, and develop natural and forest resources;
- Restore and preserve the status of ecosystems and diversity of the forest within the reserve boundary;
- Collaborate with authorized personnel to protect other resources, carry out methods for forest fire prevention, pesticide application, and preventing damaging actions in the forest;
- Set up and carry out investment projects, make annual budget estimations for the reserve's activities, and undertake a mechanism for budget imbursements at the state's regulation;
- Organize international cooperation activities through following the guideline of the authorized agency in the field;
- Set up and carry out activities and scientific research for the reserve;
- Provide services on scientific, cultural, social, and ecotourism research, followed by the government's regulation;
- Prepare periodic reports to provincial authorities and ministerial levels;
- Conduct other assignments instructed by the provincial authorities and ministry.

In terms of the relationship between the reserve's MB and the commune's PC, seemingly, least coordination in forest law enforcement is observed. More planning was needed to regulate slash and burn practices and fighting forest fires. Awareness raising should be conducted among local communities about the Forestry Law in general.

12.3.3.3 *The People's Committee of Surrounding Communes (Commune PCs)*

The commune PC has an important role in mobilizing relevant stakeholders to manage and protect forests effectively. The commune PC often organized a forest protection force to pursue and apprehend violators through forest patrols along the border of and inside the PDNR. Other activities included campaigns to prevent gold mining and devising programs to enhance local community livelihoods and relieve harvesting pressure from natural forest. The commune PC also helped to connect between specialized agencies (ie, technical experts on seedling nursery, law enforcement advisors) and local people in matters related to forest management. Nevertheless, due to insufficiency and inexperienced cooperation, the commune authorities have not undertaken their role effectively.

Ever since the model of comanagement in the PDNR was introduced, the commune PC participated from early steps of the planning process in performing activities and supervising administrative procedures. Commune authorities had an opportunity to show their roles and capacity in the local management of natural resources through supporting activities. For example, the PC of Phong My Commune had expressed their roles in preventing the deforestation and gold mining in the reserve's boundary. The commune authorities also collaborated with the reserve's MB in the planning process for forest management and forest law enforcement.

12.3.3.4 Government Mass Organizations

Organizations such as the Farmer Association, Women Union, and Youth Union have gradually performed as experts in natural resources conservation. The Farmer Association in Phong My Commune has established the Nature Conservation Association (NCA) for life with the motto "Preserving nature for the life of the future generation." This association started in 2006 with 30 members who shared the same willingness to protect the living environment.

By 2009, the NCA had gathered 90 members and diversified their activities. The initial budget was from the support of the reserve MB. This fund helped to generate household income through agriculture production, plantation of commercial tree species, and replanting bare land with green trees together with local awareness rising. The NCA took the initiative in controlling the funds as well as asking for financial support from different sources of NGOs, international funding, or funding from other government programs.

The Youth Union in Hong Van Commune (A Luoi District) had carried out many typical activities including sharing information to raise people's awareness about nature conservation, patrolling to protect the forests, and conducting research and forest inventories. Through daily activities, the Youth Union had joined nature conservation activities that actively drew participation of those people who used to practice illegal exploitation of the reserve's forests in the past.

12.3.4 Strategy of Conflict Management and Outcomes

There were some conflicts occurring after PDNR establishment. Based on the results of conflicts and the role of stakeholders' analysis, the PDNR Management Board has consulted with other stakeholders, especially representatives of communities to find out the solutions. The reason behind the conflict was restriction on access and use rights of local people after the reserve's establishment. In this sense, development of the comanagement model aimed at reducing deforestation, increasing local awareness of the importance of forests, and motivating forest conservation activities through building social capital among local communities.

12.3.4.1 Set Up Conservation Regulation in Buffer Zone

A participatory regulation on conservation management in the buffer zone of the PDNR was established and signed among the reserve MB, the commune PC, and local communities located in the buffer zone areas. The regulation was written based on legal documents. It was then read out loud and constructive comments were asked for from the 18 relevant agencies acting in natural resource areas. There were two additional meetings to collect feedback from over 100 local representatives of communities who resided in the buffer zone of the reserve.

A workshop on "establishing cooperative mechanism between local communities and the PDNR in forest protection, management and biodiversity conservation" was organized with over 30 participants. The workshop objective was to acquire the final agreement on the regulations signed by the reserve, Phong My Commune PC, and 17 relevant agencies.

Cooperative regulations between the community and the reserve were implemented together with conventions monitoring conventions built by the community through periodical meetings among the reserve, forestry agencies, and community representatives living in the buffer zone. This was intended to monitor and inform the reserve's activities and local socioeconomic development programs. Forest protection, management, and development activities were conducted once every 2 months; the first being piloted at Phong My Commune.

A network of key informants was established to provide relevant information about the status of natural resources management in the community. This network also served as a means of awareness raising between the reserve and local communities through communication on environmental education, extensions, and a policy campaign for conservation and community development. This network was first piloted in Phong My Commune and later expanded to other stakeholders and areas surrounding the PDNR such as local forest rangers, PDNR staff, and communities living in the reserve's buffer zone.

12.3.4.2 Capacity Building for Local Communities

The reserve MB has developed a thorough package for the community's capacity to sustainably manage natural forest resources. First, local communities were allocated 300 ha of forest areas under the approved district land-use plan. The allocation was properly designed in logical steps including participatory forest inventory; 5-year management planning; and preparing village regulations on forest management, protection, and resource extraction. The concept of "sustainable forest management" was specifically integrated in every step of planning, implementing, and monitoring. For example, the annual allowable cut on NTFP was determined from the forest inventory data; the scope of sanctions was based on severity and damages caused to forest resources. Such community activities helped to reduce the burden that national budget paid for forest protection annually. More importantly, the communities became real owners of the assigned forests and the sense of ownership encouraged them to promptly identify and apply measures for forest conservation.

12.3.4.3 Village Patrol Team Network

To increase the effectiveness of forest protection in the villages, we facilitated the establishment of community patrol groups (CPGs). These CPGs supported frequent coordination among stakeholders (such as PDNR officials, forest rangers, local authorities, and community representatives, among others) to enhance stewardship over forest management. As a result, 14 CPGs were established and a collaborative agreement was approved by the participating parties including local communities, a law enforcement agency (the PDNR and district forest protection department), and the commune and district PCs. In this context, forest management and protection adhered to state policy and traditional customs. During 3 years, we found that the capacity of local communities in executing institutions on forest resource management and conservation was substantially increased.

12.3.4.4 Ecotourism Operated by Local Villages

A model of ecotourism was developed within 200 ha of natural forest in the PDNR buffer zone and managed by a local ethnic minority village for income-generating opportunities. Based on the A Dong waterfall, Ha Long Village has an additional source of income for their daily food and expenses. More importantly, this source of income has greatly contributed to incentives for local people's efforts of forest protection. This successful case was later replicated in other communes in A Luoi and Huong Tra districts.

12.3.4.5 Communication Network for Conservation Activities

This network was based on the needs of local communities in exchanging good lessons and experience among various sites of the reserve. Common activities included providing technical and financial support for simple agroforestry models, a conservation workshop, and conservation awards for pupils, or local meetings for conflict resolution.

Conservation awareness was enhanced by targeting three groups (adults, youth, and children) through different structures: livelihoods, an environmental protection campaign, and learning by playing games, respectively, to create a chain effect among the communities. These activities have created positive attitudes and helped to motivate positive behaviors toward sustainable natural resource management.

12.3.4.6 Building an Ecological Village

The idea of ecological villages was to engage local villagers in natural resources management through multifarming models for livelihood improvement, income generation, and traditional custom preservation. With support from the MacArthur Foundation, two ecological villages were formulated, namely, A No Village (A Luoi District) and Ha Long Village (Phong Dien District). These villages received training and later implemented ecofriendly activities such as nursery establishment in the community, zoning forest regenerating and enriching areas, establishing forest orchards, and conserving medicinal plants through home gardens. Besides, other social benefits were achieved through using efficient wood cookstoves, raising environmental funds, and executing village regulations on forest plantation in special events.

12.3.4.7 Complementing With Other Projects and Programs

The successful models mentioned above were later applied in other projects and programs in Thua Thien Hue Province. Among them were, "building a sustainable land-use system in Nam Dong District" funded by IUCN Netherlands, "capacity building for the reserve's staff and effective law enforcement" supported by the Vietnam Conservation Fund (VCF), and "forest certification groups for smallholder plantation" by the World Bank (WB3). The models of community-based nursery garden and ecological villages were also applied in their government programs such as the afforestation program of 5 million hectares (Program 661), the poverty alleviation program (Program 135), the program on diversifying agroforestry, and the recent New Rural Development Program.

12.4 LESSONS AND CONCLUSIONS

The comanagement approach has proved successful in the case of the PDNR where conflicts between the park and local people occurred. The key principles were to bring local people into planning from the beginning, and provide them with suitable opportunities to join in the decision-making process. From the abovementioned interventions we observed the following lessons.

First, the people-centered approach was very crucial in any development intervention. To achieve real change, local resource users and relevant stakeholders must ensure meaningful participation so that they can achieve actual ownership of the change and can successfully resolve the conflicts.

Second, the best way to introduce sustainable conservation practices was the "learning-by-doing" approach. Resource management strategies should be built upon the knowledge gained through practice and application by users and stakeholders, as well as knowledge acquired by researchers. New tools and methods were essential, and attitudes of recognition and mutual respect played a vital role in this effort.

Third, property rights were considered a long-term assurance for any investment, both financial and technical. Commencing with comanagement by resource access right was essential to providing meaningful benefits to local communities for adopting some level of responsibility for natural resource management. We found that decentralization of forest management could only be successful when property rights for local inhabitants were secure.

Fourth, successful practices of forest utilization and conservation need to be institutionalized through a legal framework. Improvements in local livelihood were essential to demonstrate success of the comanagement approach; however, long-term resolution of enlarged problems often required legislative and policy reforms at a higher level. Local-level action was vital but interventions need to be adopted at higher decision-making levels to support the growth of comanagement nationwide.

References

Ashley, H., Kenton, N., Milligan, A., 2013. Tools for Supporting Sustainable Natural Resource Management and Livelihoods. Intuitional Institute for Environment and Development, London, UK.

Grimble, R., 1998. Stakeholder methodologies in natural resource management. Socio- economic Methodologies. Best Practice Guidelines. Natural Resources Institute, Chatham, UK.

Le, T.T., Minh, T.H., Ngoc, T.Q., Dung, T.Q., Hughes, R., 2001. An Investment Plan for the Establishment of Phong Dien Nature Reserve, Thua Thien Hue Province, Vietnam. BirdLife International Vietnam Programme and the Forest Inventory and Planning Institute, Hanoi.

Mayers, J., 2005. The four Rs. Power Tools Series. International Institute for Environment and Development, London, UK.

Ngoc, T.Q., Van Vinh, N., Trai, L.T., 2002. Understanding the Impacts of Hunting on Edwards's Pheasant Lophura Edwardsi at Phong Dien Nature Reserve, Vietnam: Towards a Strategy for Managing Hunting Activities. Unpublished report to the Whitley Award Foundation for International Nature Conservation.

Tordoff, A., Timmins, R., Smith, R., Mai Ky Vinh, 2003. A Biological assessment of the Central Truong Son landscape - Central Truong Son initative report No.1. WWF Indochina, Hanoi, Vietnam.

Tordoff, A.W. (Ed.), 2002. Directory of Important Bird Areas in Vietnam: Key Sites for Conservation. BirdLife International in Indochina and the Institute of Ecology and Biological Resources, Hanoi.

Payment for Environmental Services in Lam Dong and Local Forest Governance

T.N. Thang, D.T. Duong

Hue University of Agriculture and Forestry, Hue, Vietnam

13.1 INTRODUCTION

13.1.1 Background

Lam Dong is a province in the central highlands, which has a natural area of 977.219 ha including the two main parts Di Linh and Lam Vien (Langbiang), with a height of 300–1800 m. Lam Dong has 10 districts and 2 cities (Bao Loc and Da Lat). Lam Dong has a rich source of natural resources including forest and mines that are the spearhead of the province in addition to tourist advantages. Lam Dong has a large forest area occupying 38.26% of its total area with cool weather all year round, a lot of rain, a short dry season, low evaporation, no typhoon threat, and favorable conditions in economic development as well as forest management.

Lam Dong has 1,234,559 people with an average population density of 126 people/km². The two cities of Bao Loc and Da Lat have an average density of 661 and 543 people/km². This province has 50,000 ethnic minority households with 256,985 people (22% of the total population). Lam Dong has 40 ethnic minorities of which the highest is Kinh (77%), K'Ho (12%), Mạ (2.5%), Nùng (2%), Tày (2%), Hoa (1.5%), Chu-ru (1.5%), and other with lower than 1% living in the remote area in the north and northwest parts of the province. The life of ethnic minorities is hard with low infrastructure development. There are 6513 poor households and 30,457 people (12% of the total ethnic people of the province) who lack land for production.

13.1.2 Forest Status in Lam Dong

The total forestland of the province is 591,476 ha (Lam Dong, 2013) of which the special use forest is 84,153 ha (14.2%), protection forest is 172,800 ha (29.2%), and production forest is 334,523 ha (56.6%). Total forest area is 576,192 ha with 59.5% of forest cover of which natural forest is 467,099 ha, plantation forest is 109,093 ha, and barren land is 15,284 ha (see Table 13.1).

TABLE 13.1   Forest Area in Lam Dong Province Following Categories and Types

Forest Categories	Forestland (ha)		Barren Land (ha)
	Natural Forest	Plantation Forest	
Special use forest	80,780	1428	1945
Protection forest	146,375	24,932	1493
Production forest	239,944	82,733	11,846

Data from Provincial forest protection department (2013).

The decentralization process of the forest management in Lam Dong mostly involves the type of forest contract for protection. Presently the contracted area is about 376,136 ha for 22,854 households (of which 17,662 households are ethnic minorities occupying 77.3% of the land). The contract of forest for protection of the province gained a lot of agreement and support from local people to explore the benefit from forest environmental services, reduce government expenditures, and improve the awareness of local people in dealing with the benefits and livelihoods of local people and forest protection responsibility.

The forest contract in Lam Dong now is of two types: contract forest with monthly payment and forest contract for plantation benefit.

- Contract forest with monthly payment:

    The total forest and forestland contracted for protection in the province is 368,609 ha (64% of the forestland of the province); forest owners contract with 19,995 households (of which 16,386 HH are ethnic minority). In this category, there are several sources of funding:
    - The provincial budget: 24,401 ha
    - The Flitch project: 23,237 ha
    - Payment for Environmental Services (PES) policy: 320,970 ha
- Forest contract for plantation benefit:

    The total area contracted is 5665 ha for 2757 HH, of which 1566 households are Kinh people, and 1176 ethnic minority HHs. In this category, there are several subcategories of policy and regulations that the contract in the province follows:
    - Degree 135/2005/NĐ-CP: 3252 ha
    - Decision 178/QĐ-TTg and Degree 01/CP: 1409 ha
    - Forest contract following the policy of 30a: 1004 ha

In addition, as a test of the decentralisation process, to 2013 there were 15 local communities in 8 communes and 5 districts being allocated with a total area of 2,015.29 ha. However, as it is in a very small scope, this study does not take into consideration this allocation area for evaluating forest governance.

13.2 STUDY OBJECTIVES

The main study objective is to assess the current forest protection and management and local forest governance along with the PES program in the local context. What are the current institutions involved? How is the forest governed at the local level and how effective is it for the current PES program in Lam Dong Province? The result of this study would provide the foundation and baseline information to the policy makers in issuing suitable policies and regulations for sustainable forest management and livelihood improvement for local people.

13.3 METHODOLOGY

To make it completely a participatory process, we start by forming a working group that includes members of the provincial and district departments and commune officers. Representatives from the provincial levels to the grassroots levels of Lam Dong with support from the research group implemented the whole study process.

The criteria for selecting two communes are: (1) they should contain a high percentage of K'Ho households, who have a high dependency on the forest for their livelihoods; (2) these communes still possess a high percentage of natural forest cover; and (3) households in the villages have good access both to forest and markets.

Several preliminary reconnaissance visits were taken to the two study communes to hold informal interviews with key informants and establish a sampling design. Interviews were held with commune leaders and officers, village headmen, and key K'Ho village elders. Preliminary discussions were held about forest governance in the local context and current forest protection management and utilization and forest dependency data were collected and analyzed. We used clear criteria for choosing key informants such as (1) in-depth understanding of the issue of local forest governance; (2) being in charge of management and technical jobs that are relevant to the assessment; and (3) being connected, working directly or indirectly with those who are related to forest protection activities.

Semistructured interviews to investigate historical and current patterns of K'Ho forest use and forest governance were conducted with a total of 20 key informants. Key informants were village headmen, village elders, commune leaders, and representatives of the DARD, DONRE, and FPD.

The results revealed that there was a high degree of similarity across K'Ho villages within a commune and across the two study communes. Based on the preliminary survey, we concluded that one village survey in each commune would be sufficiently representative of K'Ho villages in Lam Dong. Within each commune, one K'Ho village, which has contracted forest, was randomly chosen for surveys. Thus, in this study we aggregate the results from two villages to make it representative of K'Ho people. From the list of households in the village, we randomly chose the households for questionnaire interview. A total of 96 households were surveyed out of the total 148 households in the three villages.

The criteria for choosing households for interview were (1) receiving allocated forest; (2) high level of dependency on forest resources; (3) households are categorized by level of income (poor, average, above average); and (4) gender (to ensure that at least one-third of the participants are women).

Finally, a group discussion was held in each village at the end of the fieldwork to discuss the information that had been collected and to build consensus on its reliability. The criteria for selection among the individual households, members of the community, and other individuals were (1) in-depth understanding of forest allocation and contract; (2) village chief, elder, village patriarch with good knowledge and experience; (3) families with high level of dependency on forest resources who receive/are allocated forest; and (4) people who collect forest products and/or receive allocated forest. In Bao Thuan Commune, some forestland is allocated to the community rather than to individual households. Some representatives of the community were also invited.

There are several ways of data collection in the study area:

– Information on local forest governance by group members collected during key information interview and group discussion regarding all matters related to forest governance. This also includes the information related to the nonrecipient villages.
– Data are collected through a household questionnaire on forest use and forest protection, other livelihood activities, changes in forest management, and protection before and after forest contract.

The qualitative analysis was used for analyzing the forest governance associated with the process and de facto rights generated and practiced by local people.

13.3.1 Study Area

Di Linh District implemented the UN REDD Vietnam program in Phase 1 (2009–11) so there have been many activities related to forest preservation and protection. Most forest areas in the district are planted forest (pine trees), which are managed and developed by state-owned forest owners as forestry companies. Two major forest owners in the district are Bao Thuan Forestry Company (owning 18,913.44 ha) and Di Linh Forestry Company (owning 29,971 ha, of which 27,051 ha is old-growth forest). Both companies are contracting forests for protection to local people to increase the effectiveness of forest management and protection as well as to enhance local people's livelihoods. The main capital for forest management and protection contract is from the provincial budget and the PES, with some financial assistance from different projects (Fig. 13.1).

Bao Thuan has a very large planted and productive forest, managed by One Member Bao Thuan Forestry Company, Ltd. Even though the forest area is large, up to now the company has only been able to allocate almost 4000 ha to 180 households for protection.

About 88% to 89% of the forest area in Lac Duong District is watershed forest, playing an important role in preserving the water source for hydroelectric plants located in the area. While this district did not participate in Phase 1 of the UN REDD Vietnam program, activities of Phase 2 of the UN REDD Vietnam program were implemented here from 2013 to 2015. Most forest area in the district is special use old-growth forest and protective forest. Two major forest owners in the district are Bidoup-Nui Ba Natural Reserve and the Management Board of Da Nhim Productive Forest. These two organizations have allocated forest to local people for management and protection.

In addition, there were private forest owners who receive allocated forest or rent forest for productive and tourism activities such as the Lac Duong District Police, the District Military Steering Committee, the Provincial Military Command, Ward 12 Police, Da Nhim Commune Police, and others.

Da Chais Commune has a very large old-growth forest area. The Bidoup-Nui Ba Natural Reserve (special use forest) and the Management Board of Protective Forest Da

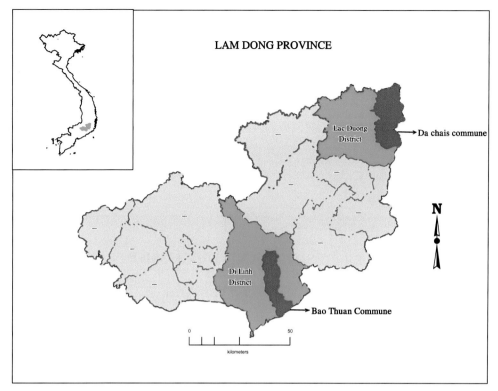

FIG. 13.1   Study area in Lam Dong Province.

Nhim (special use forest) manage a large part of this forest area. These organizations allocate a relatively large area of forest to local people for management and protection. The total area these two organizations allocate for management and protection in Da Chais is 17,514 ha.

The similar features of the two areas where data were collected are a high proportion of ethnic minority groups and quite large coverage of old-growth forest. However, the two areas also have very distinctive physical and socioeconomic characteristics.

Lac Duong District is very close to Da Lat City; thus it enjoys a close connection and good trading practices, which leads to better economic status. Consequently, the management and protection of forest resources benefit from certain advantages as well as the cooperation with relevant provincial agencies. However, this also creates pressure on the forest resources of the region. With diverse resources, convenient transportation, and good road conditions, this region is under pressure from illegal exploitation, hunting, and transport of forest products. Di Linh District is quite far from Da Lat City; thus road conditions make it more difficult to transport forest products.

The number of households selected for the survey and group discussion in each village was decided based on the ratio of households in each site as presented in Table 13.2.

TABLE 13.2   Some Information About the Villages in the Study Area

	Commune	Village	No. of Interviewed HH	Note
1	Bao Thuan Commune	Hang Pơr	15	Poor village
		Kla Tô Kreng	15	Above average
		Kla TằnGu		This village has allocated natural forest to the community. Four to five people were invited to the discussion to provide different perspectives on forest allocation and forest contracts
1	Đa Chais Commune[a]	Village 2	15	Above average village
		Village 3	15	Poor village

[a] Group interviews in Da Chais Commune were conducted with people in Village 2 and Village 3 and no other participants from other villages.

13.4 RESULTS AND DISCUSSION

13.4.1 The Current Status of Local Forest Governance

The current status of local forest governance of the study area focuses on forest contract/allocation to improve local people's livelihoods. The information was derived from secondary data of the commune People's Committees, the district forest management department, and forest owners in combination with primary data collected from in-depth interviews with relevant stakeholders, households, and group discussions. Key components for forest governance in "forest allocation to improve local people's livelihoods" are demonstrated below.

13.4.1.1 Forest Condition Before Contract

Table 13.3 shows indicators of forest condition before allocation in terms of land area categorized by functions, types, and status. It also includes the number of groups and households that receive/contract forest for protection and management in the two areas with the total area of land, forest, and bare land.

The majority of productive and protective forest in Bao Thuan Commune, Di Linh, has low timber volume, ranging from 21 to $110\,m^3/ha$, while the natural forest area in Da Chais, Lac Duong, has very high timber volume, ranging from 218 to $235\,m^3/ha$. Consequently, the average area allocated to each household in Da Chais is quite large, about 28–48 ha/household. Meanwhile, the average area contracted to each household in Bao Thuan, Di Linh, only ranges from 2.5 to 21 ha. This difference creates a gap in income from forest protection and management activities of local people in the two study areas.

13.4.1.2 Rights and Responsibilities in Allocation/Contract Forest for Protection and Management

Income structure from livelihood activities could show the contribution of livelihood activities in the total income of local people (Table 13.4).

In both Bao Thuan and Da Chais, the biggest source of income comes from growing and harvesting coffee, which makes up 55% of the total income of people in Bao Thuan and almost

TABLE 13.3   Forest Condition Before Contract

			Area (ha)			
	Forest Owners	**Group/Households**	**Total**	**Forest**	**Bare Land**	**Volume (m³)**
Di Linh (Bao Thuan Commune)	Bao Thuan One Member Forestry Company, Ltd	17 groups/180 households	3849	3779	70	83,426
	Community forest of Kla Tangu Village	4 groups/196 households	500	500	0	55,368
Lac Duong (Da Chais Commune)	Da Nhim Protective Forest Mang. Board	6/112 households	3240	3240	0	709,447
	Bidoup-Nui Ba National Park	30 groups/308 households	14,274	12,847	0	3,357,741
Total		57 groups/786 households	26,363	24,867	70	4,205,982

Source: District forest protection departments (2013).

TABLE 13.4   Income From Forest Allocation/Contract and Contribution to Total Income

	Average Income in Bao Thuan and Da Chais		Bao Thuan		Da Chais	
Sources of Income	**Income**	**Ratio (%)**	**Income**	**Ratio (%)**	**Income**	**Ratio (%)**
Forest allocation	8,664,440	15.6	6,769,379	11.1	10,559,500	21.0
Animal farming	2,000,000	3.6	–	–	4,000,000	7.9
Coffee	23,762,011	42.7	33,689,655	55.3	13,834,366	27.5
Rice/vegetables	4,581,810	8.2	4,163,619	6.8	5,000,000	9.9
Business	1,910,204	3.4	–	–	3,820,408	7.6
Salary	5,051,667	9.1	5,800,000	9.5	4,303,333	8.5
Work as hired labor	9,663,753	17.4	10,486,364	17.2	8,841,143	17.6
Total income	55,633,884	100.0	60,909,017	100.0	50,358,750	100.0

Source: Household interviews (2013).

28% of those in Da Chais. Income from forest management and protection in Bao Thuan only makes up 11% of total income, and 21% of that in Da Chai. This is a significant source of income here. Working as hired labor makes up about 17.5% of the income in both areas. As a whole, people in Bao Thuan earn higher incomes than those in Da Chais.

Whether payment for forest management and protection is made on time is an important indicator to demonstrate local people's rights in forest contract/allocation.

In Da Chais, during the group discussions, local people raised a lot of concerns about late payment for their forest management and protection efforts. This reflection is expressed quite

TABLE 13.5   Time of Payment for Forest Management and Protection

Commune	On-Time Payment		Late Payment	
	No. of People	Ratio (%)	No. of People	Ratio (%)
Bao Thuan—Di Linh district	23	79	6	21
Da Chais—Lac Duong district	14	43.7%	18	56.3%

Source: Household interviews (2013).

clearly in Table 13.5, as 56.3% people indicated that payment is made later than it should be. Local forest governance should pay attention to this.

During our data collection, there was no adjustment in payment for households or groups of households because there was no change in area or other reasons such as forest fires. However, this indicator could be tracked through time (ie, once per year) and this job requires coordination among forest owners who conduct periodic inspection, households who receive contracted forest, and the fund for forest environmental services. In fact, adjustment in payment for households/groups of households can be easily gathered from forest owners.

In reality, some households have violated forest management and protection activities and their payment in the contract was deducted. However, if only this sanction was applied, effectiveness and level of deterrence is not high and forest resources face a high risk of exploitation as profit that could be made from forest exploitation/violation is much higher than payment for the management contract. Some households are willing to have their payment deducted to earn a higher income from illegal activities that violates the forest management regulations.

13.4.2 Effectiveness of Forest Management and Protection After Contract

There are big differences in forms of violations and severity of forest damage in the two areas. In Di Linh, illegal deforestation happens really often (128 cases) compared to Lac Duong (28 cases). In Di Linh, deforestation mainly happens because of a switch in purpose of land usage (ie, from forestland to agricultural land and land for growing long-term industrial trees), while in Lac Duong most deforestation happens because of logging and animal hunting. Therefore, sanctions and fines in Lac Duong are much more severe than those in Di Linh. Similarly, the number of illegal trading and transport of forest products in Lac Duong is very high (116 cases) compared to Di Linh (25 cases). Convenient road conditions as well as close proximity to big centers for consumption (eg, Da Lat City) are the main reason for this problem (confirmed by group meetings) (Table 13.6).

Fire does happen but not frequently; only one fire took place in Bao Thuan One Member Forestry Company, Ltd., and damaged 15 ha, but it was an area covered with grass, not forest. People cooking in the forest caused the fire (report of the district forest protection office).

Due to different physical conditions in the two regions, forest fire prevention is quite different. In Lac Duong, most forest is green, natural forest so there is a low risk of fire. On the contrary, most of forest in Di Linh consists of human-grown pine trees so the severity and extent of risk of fire is much higher than other areas.

Information about an indicator of increased timber volume was not gathered as there is no assessment of timber volume before and after contract because this job would require very

TABLE 13.6 Violations and Damages to Forest Resources in the Region

Offense	Number of Cases		Damaged Area (ha)		Damage Forest Products (m³)		Extent of Damage (Thousand Dong)	
	2012	2013 (6 months)	2012	2013 (6 months)	2012	2013 (6 months)	2012	2013 (6 months)
Deforestation	158.0	72.0	26.9	14.4	374.1	152.0	1,749,200	318,000
Illegal exploitation of forest products	25.0	8.0			124.4	23.6	121,000	410,000
Violations of forest fire prevention, causing fire	5.0	4.0	3.0	8.4				
Illegal transport of forest products	12.0	4.0			9.3	17.2	76,000	96,000
Illegal trading, storing, processing forest products	13.0	2.0			83.3	2.0	15,000	
Encroachment of forestry land	6.0	3.0	1.4	3.0			6000	11,000
Di Linh total	219.0	93.0	31.3	25.8	591.1	194.8	1,967,200	835,000
Deforestation	27.0	6.0	8.9	0.5			1,175,081	85,903
Illegal exploitation of forest products	132.0	37.0	236.7	118.6				
Violations of forest fire prevention	1.0							
Violation of timber processing regulation	1.0							
Illegal trading and storing forest products	116.0	59.0	58.5	38.0				
Encroachment of forestry land	4.0	1.0						
Lac Duong total	281	103	304.1	157.2	–	–	1,175,081	85,903
Total	500	196	31.3	25,797	591,086	194,778	3,142,281	920,903

Source: District forest protection office (2013).

high expenditures of human and material resources. In addition, the fact that boundaries among groups/households are not clear also causes difficulties for the assessment of this indicator. This shortcoming needs to be considered and overcome. To assess the effectiveness of forest governance, especially productive forest, timber volume is a key indicator to indicate the effectiveness of forest resources management and growth. In particular in the coming time, when activities of the REDD+ program are implemented in the two areas, timber volume assessment as well as annual growth (biomass) of forest contracted for protection and management in the two areas must be integrated and implemented so that the effectiveness of forest governance is identified as well as creating a more accurate payment system for the people participating in forest protection.

13.4.3 Forest Governance Status From Local People's Perspectives

The interviews with households and the group discussions were conducted with the purpose of enhancing local people's participation in forest governance. So, in addition to collecting some information for the indicators and cross-checking information collected elsewhere, they also provided an overview of factors that influence local people's participation in forest governance.

There are some differences in forest governance in the two areas. These differences are due to variations in how forest owners proceed and approach local people, as well as natural and socioeconomic differences in each area. However, fundamentally both the testing sites share some common features.

- The liaison between forest owners and local people mostly is key contacts such as village chiefs, head of the contract group, and commune People's Commune. Local people often receive information about the forest contract for protection and management through these contacts.
- From the discussion, the criteria for a household to receive a contract of forest protection are as follows:
  - Having an available laborer.
  - Do not violate and practice illegal forest resources extraction.
  - Priority is given to poor households or ethnic minority households.
  - Having a strong desire to participate in forest management and protection.

In Bao Thuan, local people through group discussions requested to add three more criteria for choosing households to participate in forest management and protection.

- Maintain fairness by rotation so all households can participate in forest management and protection.
- The community should be consulted in evaluating households who participate in forest management and protection; priority should be given to households with high commitment to forest protection.
- Community evaluation for households with a forest contract should be carried out annually.

This shows that at the moment local people have not found the contracting process to be fair and they want to participate in the whole forest governance process in their areas including selecting who participates in the contract.

- In Lac Duong all households received contracts so they did not show much interest in the duration of the contract. However, the number of households receiving contracts is increasing so the amount of money each household receives from forest protection is decreasing. On the other hand, in Di Linh only some households received contracts and other households must wait for an uncertain length of time. Local people expressed that each contract should last for 2 years instead of 5 years so other households can also participate and get the benefit from a forest contract.

- One obvious problem here is that the forest contract does not encourage participation from households who have above average income, as they may not be interested in forest protection and management or the payment from the contract is not significant to them. This is a weakness of the contract process, which excludes households with above average incomes from local forest governance.
- Local people's goal and strongest interest in a forest contract for protection is to increase income. More than half (58%) of the people interviewed said the current payment rate for forest management and protection is too low, 37% thought it is reasonable, and 5% considered it too high. On average, income from forest protection and management contributes 15.6% to the total income of households with forest contracts (21% in Lac Duong and 11% in Di Linh). Therefore, this is a significant source of income for local people, especially poor households.
- Local people have not been able to really participate in the contract as they play a relatively passive role. They have not discussed how to identify their roles, responsibilities, and rights but mostly followed other people's instructions. The majority of local people (73%) believed that the procedures for them to receive a contract were simple and convenient. They just needed to submit a copy of their IDs, sign the contract, and then they received quarterly payment. Most people who have signed the contract do not know the terms written in the contract, and do not even keep a copy of the contract. This should be noted in the contracting process because it shows that local people have not truly participated in the forest governance process.
- Most of the households do not know clearly where their contracted forest is, but only know where it is located within a group. This shows the important role of groups and the team leader of a group. However, this also signifies the lack of active participation and clear understanding of people's rights and responsibilities, which leads to difficulties in personal accountability when violations of forest management and protection happen.
- Local people are not very interested in participating in the forest protection process. They do what they are assigned; mostly jobs given by forest owners. They consider themselves hired workers and receive payment rather than taking control of forest protection and management. Some households in Bao Thuan rarely participate in forest protection activities even though they have contracts.
- Agencies and individuals whose roles are important in local people's contracts are forest owners, village chiefs, and heads of contract groups and commune People's Committee. The most supportive agencies before and after the contracting processes are forest owners, village chiefs, local forest rangers, and commune People's Committee. If local people need to send feedback about forest protection and management activities, they would meet the head of their group, and then the village chief. Afterwards, the village chief would address the problem to the commune People's Committee and the forest owner.
- During discussion, local people seem to understand quite well the rights, benefits, and responsibilities of their households with contracts. However, they mostly follow instructions and requests of forest owners, the village chiefs, and the head of their groups in forest protection activities.
- In forest protection and management activities, local people have access to information about fire prevention and firefighting, prevention of slash and burn farming, exploitation

of forest products, and hunting and transport of wild animals. Most local people have access to information spread by the commune People's committee, forest rangers, and forest owners.

- The main forest protection and management activity most households are currently doing is patrolling. They divide the tasks among groups and teams, and patrol periodically. Forest development activities have not really been paid attention to.
- Contract duration: different households have different contract duration. In Da Chais, contracts last for 1 year, while in Bao Thuan, contracts used to last for 2 years and recently they last for 5 years. Local people indicated that the duration of protection contract of 2 years is reasonable (Bao Thuan), because contracts should be rotated so other households could in turn receive forest to protect and manage. In Da Chais, local people take the activity for granted and they do not pay attention to the duration (as forest area is large and they know that they will be contracted every year). On the contrary, in Bao Thuan, there are 200 households but only 20 households receive the contracts, so local people really care about the duration of contracts. With the current mode of rotation, as the contracts last for 5 years, some households must wait for a really long time before they could receive the contract.
- For households with allocated community forest in Bao Thuan (Kla TầnGu village), local people have started to understand the benefits of forest allocation. Currently, forest is allocated for their community for 50 years, and in addition to the benefit from timber and NTFPs, they could benefit from PES. However, they have not realized the impact of forest development and have been only focusing on protecting the area of forest already available.
- Having understood the differences between contract and allocation, the local people were asked which mode of engagement they preferred. There is a remarkable difference between the choices of people in Di Linh and those in Bao Thuan. About 65% of people in Bao Thuan wished to be allocated forest while only 19% of people in Da Chais did. This is understandable given the fact that most forest area in Da Chais is special use and protection forest while forest in Di Linh is production forest. In addition, contracts for a large area of forest could generate a significant amount of income for people in Da Chais and that motivates them to receive contracted forest.
- When answering the question above, some people expressed concerns over their ability to protect allocated forest. Some households preferred contracts because allocated forest is often poor while local people do not have capital to invest, as well as the ability to protect the forest. When asked whether they would be interested if allocated forest came with seeds and technical assistance, they were enthusiastic and stated that they would develop plantation forest to bring benefits for their families and society.
- In addition, local people also presented some comments on ways to improve forest protection and management activities from their perspective:
  - Support in making a detailed map of the allocated and contracted forest so the people understand well their areas and their tasks.
  - Professional, technical training in forest protection and care.
  - Provide safety equipment and fire prevention equipment for local people.
  - Technical training in seedlings, nursery, forest recovery, regeneration, and development.
  - Provide herbal plants, forest products besides timber, and rare timber species for the people to grow to improve forest quality and people's income.

13.4.4 Challenges in Local Forest Governance

Two general issues in local forest governance that were identified are (1) decision making and a mechanism to make decisions in forest management and protection and (2) whether current forest governance really brings benefits to local people. However, there are still gaps and shortcomings that need to be addressed to ensure effectiveness and sustainability of the current forest governance process.

13.4.4.1 Benefit-Sharing Mechanism

Currently, the value of contracts is withdrawn from the forest environmental fee and allocated in the following way: 10% for the staffs of the fund for forest environmental services, 9% for forest owners, and 81% for people with contracts. Added value from timber goes to forest owners exclusively.

- Local people with contracts receive 81% of the fee for forest environmental services but they are not responsible when deforestation happens due to fire, illegal exploitation, or switch of usage, unless they themselves cause the damage by doing those activities. They do not benefit from the added value of protecting and managing the contracted forest. The current rate of payment (8.6 million VND/household/year) is not high enough to incentivize people to participate.
- Forest owners receive 9% of the environmental services fee and the added value (timber) of forest resources.
- Other agencies such as forest rangers and commune People's Committee do not benefit from the governance activities.
- With the current forest contracts, local people do not have the right to make changes to the forest, even improvement activities (Fig. 13.2).

We concluded that this benefit-sharing mechanism affects and reduces the incentives and participation of local people and related stakeholders in forest resources management and protection.

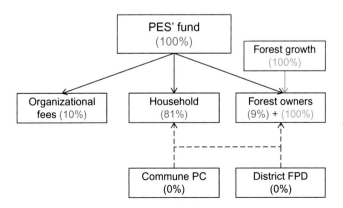

FIG. 13.2 Current benefit-sharing mechanism and stakeholder's incentives.

13.4.4.2 *Actual Participation of Various Stakeholders in Forest Governance*

- The commune People's Committee (or the commune forest management department) only participates as a local administrative agency and fulfills their management responsibility without any motivation to participate actively in forest protection and management.
- The district forest management department is the government agency to take charge of forest protection and ensure that the law is followed when it comes to forest protection, management, and forest products management. They also coordinate with and support forest owners and commune People's Committees.
- Local people who participate in forest contracts in reality are hired laborers who receive low payment, even though technically they receive 81% of the total fee paid for environmental services. In addition, the contract is mostly a formality and no clear legal regulations identify their specific roles and responsibilities in the management. They do not benefit from added values from the forest (timber growth value). The fact that local people are lacking productive land, together with the high value of industrial trees, creates pressure on conversion of land use or encroaching forestland to plantation forest. For those reasons, the effectiveness of forest management in the area is quite low, and deforestation and degradation of forest products continues to take place.
- Local people have not truly participated in local forest governance. They play a passive role in the whole contracting process for forest protection. For the most part, they are not informed and able to contribute any thoughts to the process of distributing and receiving contract and implementation of forest protection activities. This one-way information sharing and decision making leads to a high risk of forest degradation.
- Information collected shows that local people have not participated in the development of a forest protection and management plan for contracting, except for submitting a copy of their IDs and signing their contracts. They considered the procedures very simple; however, this simplicity might present potential risks for local people because they did not keep a copy of the contract and did not understand what their rights and responsibilities are besides patrolling as forest owners requested.

In fact, passive participation is a consequence of the first issue; namely, the benefit-sharing mechanism. As the mechanism is not clearly defined and stakeholders do not see their interests, their participation will be a formality and will not be effective in forest resources management and protection.

13.5 CONCLUSIONS AND RECOMMENDATIONS

13.5.1 Conclusions

13.5.1.1 *Status of Local Forest Governance*

The two areas have different conditions and forest owners have actively participated in the forest contracting process to enhance the participation of the local community in forest protection and management as well as to improve local people's livelihoods.

The current status of local forest management is shown quite clearly though.

13.5.1.2 *Roles, Functions, and Responsibilities of Stakeholders*

The commune People's Committee is a government administrative institution that supports forest owners in managing, protecting, and developing forest resources. The commune forest management department is directly involved in activities jointly conducted by forest owners and local people, with specialized staffs who receive allowances from the state budget and the fund for forest environment services. This institution plays an intermediary role, receiving and sharing information related to forest resource protection and management, organizing communication activities to raise local people's awareness on forest protection and management, prevention of forest fires, patrolling, supervising, and developing forest resources.

Forest owners receive forest resources assigned by the state, playing an important role and bearing responsibility for the protection, development, and management of forest resources. They give contracted forest to households/groups of households for protection and management to reduce pressure on forest resources and enhance local people's participation in protection and management and improve their livelihoods.

The district forest management department carries out the state regulations on forest protection, ensuring the law on forest protection and development and forest products management is followed. They support forest owners to fulfill their tasks.

Local people receiving forest contracts for protection participate and benefit from patrolling and supervising the forest resources. However, they have not really participated in the local forest governance process. They are quite passive in the whole contracting process. They almost never participate and raise their ideas during the contracting process or in distributing and receiving forestland.

Through the discussion with local people, it appears that the roles of the village head and head of the group are really important in addition to the role of the local forest ranger. They are the connection between the forest owners, local authorities, and functional agencies to local people and communities. It is really important to strengthen their roles, responsibilities, and rights toward sustainable forest management.

13.5.1.3 *The Benefit-Sharing Mechanism and Participation of Stakeholders*

In the current forest contracting mechanism, local people who enter a contract benefit the most from PES as they receive 81% of the total amount. However, this source of income only contributes 15.6% to household total income. This shows that forest protection has not yet been an attractive option for local people.

There needs to be a support mechanism so local people could benefit more from their participation in forest protection. A benefit mechanism from forest protection (ie, added value, in particular timber growth or REDD+) will be an incentive for people to participate actively and sustainably in forest protection.

In addition, other stakeholders such as the district forest management department and the commune People's Committee should also participate and enjoy in this benefit mechanism to ensure the sustainability of forest resources management and protection.

Currently, households with above average income are not motivated to participate in the forest protection contract because either they are not interested or the payment is not significant for them. This is a shortcoming of the forest protection and management contract, because it unintentionally excludes households with above average income from forest governance.

Local people, to some extent, participated in forest protection and management activity. They have proposed some feedback to the selection of households for the contract, identified an appropriate duration of the contract, and wished to receive forest to use for the long term.

13.5.2 Recommendations

The local forest governance activity should implement the following changes to improve the participation and sustainable management of forest resources.

- Enhance the role of the stakeholders in the forest governance process, especially local people and communities, village heads, team leaders of the household groups, and local forest rangers.
- Develop an information-sharing mechanism about the forest governance process so that all related stakeholders know clearly their rights and duties in forest protection and management.
- Develop a clear benefit-sharing mechanism so that all stakeholders who participate in local forest governance activity can benefit from it so that their incentives and responsibilities are clear and can contribute to the sustainability of forest management.
- Promote local people's participation through awareness raising, capacity building activities, and other economic incentives.

References

Decision 178/2001/QD-TTg, 2001. Government Decision on the benefits and obligations of households and individuals assigned, leased or contracted forests and forestry land.
Decision 18/QĐ-UBND, 2013. Lam Dong provincial Decision on approving the plan on forest protection and management from 2011–2020.
Decree 01/CP/1995/TTg, 1995. Prime Minister Decree on guiding land allocation through contracts for agriculture, forestry, and aquaculture purposes.
Decree 135/2005/ND-CP, 2005. Government Decree on contractual assignment of agricultural land, production forestland and land with water surface for aquaculture in state-run agricultural farms and forestry farms.
Resolution 30a/2008/NQ-CP, 2012. Government resolution on the Programme of Rapid and Sustainable Poverty Reduction in poor districts and the National Target Programme of Sustainable Poverty Reduction, amendments and supplements to policies for near poor households.

Is Vietnam Legally Set for REDD+?

S. Sharma*, G. Shivakoti[†,‡], T.N. Thang[§], N.T. Dung[§]

*WWF Nepal, Kathmandu, Nepal †The University of Tokyo, Tokyo, Japan ‡Asian Institute of Technology, Bangkok, Thailand §Hue University of Agriculture and Forestry, Hue, Vietnam

14.1 INTRODUCTION

14.1.1 Background

Forest management in Vietnam is based on forest policy and legislation with a general subscription to sustainable forest management and environmental protection. Management of forest is important for sustainable development in Vietnam. However, Vietnam lost approximately half of its forest during 1943–90 (Nguyen, 2013; FAO, 2010), which confers on Vietnam the distinction of having the second highest deforestation rate in the world; further aggravated by climate change. The 13th Conference of the Parties (COP 13) identified Vietnam as one of the five countries to be affected by climate change. The consequences of climate change have been prominent through reduced forest and agriculture productivity and decline in quantity and quality of water resources; simultaneously intimidating economic gains mostly gained through the *doi moi* policy.

Concrete ideas are undertaken to adapt and mitigate climate change effects. Vietnam is among the 13 pilot countries that have been chosen by the UN-REDD to participate in future REDD+ mechanisms. Simultaneously, Vietnam has also sent a REDD+ Readiness Preparation Proposal (RPP) to the Forest Carbon Partnership Facility (FCPF). Since then, Vietnam has developed strategies and issued action plans to address multiple causes and drivers for deforestation and degradation coupled with maintaining social and environment benefits. But much of the literature suggests the process of policies and strategies formations do not consider local needs and aspirations. Prior to implementing REDD+, it is important to know the existing legal policies in terms of REDD+ benefit sharing and creating scope for better livelihood through forest restoration. This is the initial stage for REDD+ planning, beyond which policy considerations cannot be easily incorporated; a study merely focused on the possibilities of REDD+ on the basis of existing policy provisions is crucial to look for areas that need extra work.

The major objective of this study is to analyze and enlist legal provisions that are relevant to REDD+ interventions in Vietnam. Ultimately, the chapter anticipates drawing recommendations to enhance Vietnam in strengthening the institutional and legal capacity to implement REDD+. As the chapter intends to review existing forest policies and their impact on local communities, we focused our desk policy review. Policies relevant to REDD+ governance are reviewed. This study is part of the project "Bridging Policy Practice Gap in the Effective Implementation of REDD+ Programs in SE Asia: Collaborative Learning among Indonesia, Thailand and Vietnam," with financial and technical support from the Toyota Foundation from Nov. 2013 to Nov. 2014.

14.1.2 Vietnam's Governance Situation for REDD+ and Development Goals

There are two parallel systems in state government in Vietnam. The constitutional system has a mandate of formulating and passing laws and resolutions from the state to local levels; namely, the National Assembly-Provincial Council-Community Council. The executive system is responsible for executing laws and policies (decisions, decrees, etc.) that include the government, provincial committee, district committee, and commune committee. Vietnam is pertinent to REDD+ preparedness as it regulates the introductory commands of the players who design and instrument REDD+ at the national, provincial, and local level. The country is managed at central, provincial, district, and communal levels. The National Assembly is the highest organization at the state level that approves laws. The government issues decrees that explain laws in detail. The provincial people's committee (PPC) is a locally elected body active at the provincial level. The district people's council oversees law execution at the district level often is backstopped by professional agencies and the Commune People's Committee (CPC) at the community level. Though Vietnam's state power resides centrally, the laws are enforced nationwide. Nonetheless, definite competences are decentralized to local government that have competency and authority to implement polices within a given jurisdiction. REDD+ laws and strategies are required to be enacted centrally but are also vital to improve local capacity to foresee effective REDD+ implementation in Vietnam.

The Cancun Agreements provides an opportunity for Vietnam to combine climate change mitigation together with achieving both national and global development goals. Vietnam is praised for its *doi moi* economic policy, but challenges still persist in respect to rural and mountainous forest-dependent ethnic communities who still live in poverty. Much of Vietnam's economic growth is only possible due to exploitation of natural resources and the country now needs to revisit polices to ascertain that natural resources will be harvested below the regeneration level. According to Bojo (2011), "Vietnam's growth will need to assign strong property rights, transaction rules, conflict resolution mechanism, community benefit sharing schemes, transparent government and people's participation in decision making."

Furthermore, the government has initiated some sustainable approaches explicitly through Vietnam Agenda 21 and Socioeconomic Development Strategy (SEDS). Vietnam Agenda 21 stands at three pillars: economic development, social equity, and environmental protection; dimensions for sustainability. SEDS is the economic road map for advancing as an industry-based country by 2020. It is determined to increase forest cover up to 45% along with social equity and development. Remarkably, climate change response is a national priority in Vietnam. Socioeconomic development plans (SEDP) have been developed to implement

SEDS with natural resource management as the primary focus. Besides, SEDP has set aside 42.5% of forest cover targets by 2015. Referring to these national targets, REDD+ is in an advantageous situation in Vietnam.

Vietnam is a signatory to most international conventions that promote REDD+ implementations; the Cancun Agreements is one. Others include the Convention on International Trade in Endangered Species of Wild Fauna and Flora (CITES), the Convention on World Heritage, the Ramsar Convention, the United Nations Framework Convention on Climate Change (UNFCCC), the Convention on Biological Diversity (CBD), and the United Nations Convention to Combat Desertification (UNCCD). Vietnam has strengthened its relationship with the United Nations Forum on Forests (UNFF), the Food and Agriculture Organization of the United Nations (FAO) Committee on Forestry (COFO), as well as other regional organizations such as the Asia-Pacific Forestry Commission (APFC), the Asia Forest Partnership (AFP), and the International Network for Bamboo and Rattan (INBAR). Vietnam is actively involved in a number of international initiatives related to forestry and signing trade agreements with a number of forest products and international organizations and countries, such as sustainable forest management and forest certification; REDD+; Forest Law Enforcement, Governance, and Trade (FLEGT); voluntary partnership agreements (VPAs), the European Union Timber Regulation (EUTR), and the U.S. Lacey Act. Under these doctrines of legitimate expectations, these instruments can guide Vietnam into the realm of REDD+ polices.

14.1.3 Public Participation and Admission to Information

The Cancun Agreements call for the effective participation, particularly of ethnic people, in developing and mainstreaming national REDD+ polices and strategy plans. It demands transparent forest governance information on ways that REDD+ interventions and safeguards are being addressed in the policy. In Vietnam, the following commitments strengthen public participation:

(a) Politics Bureau (Resolution 41/TW-BCT) specifies the mandatory involvement of organizations and individuals in environmental protection. For this purpose, the country has provided the right to participation through the right to information.
(b) Vietnam's Law on Environmental Protection (2001) provides rights to people to actively participate in environment impact assessments in their residential areas.
(c) Vietnam's Law on Land (2013) addresses transparent land-use planning mechanisms with active public participation.
(d) Vietnam's Agenda 21 (National report at the United Nations Conference on Sustainable Development (RIO+20), 2012) provides the right of local participation in sustainable development.
(e) The country has laws related to access to information and a law on complaints and accusation.
(f) The indigenous societies are provided rights to remark on relevant polices regarding their rights and commitments.
(g) The Forest Protection and Development Law 2004 (FAO, 2005) specifies the need for forest protection planning to be approved by public. In the process, the required information is to be provided by the state.

Many of Vietnam's indigenous people are forest dependent and reside near them; they have a long-enduring relationship that has usually been recognized legally and customarily. The active involvement of these people in forest management is likely to fetch good results. In a way, the legal provisions in Vietnam have significant space to incorporate local needs and aspirations at each level, which is crucial to increase the success levels in REDD+.

14.2 IN EFFECT POLICIES WITH IMPLICATION TO REDD+

This section identifies legislative provisions and mandates pertinent to REDD+. The major formal legislation governing future REDD+ programs includes in following areas: (a) climate change strategies; (b) land-use planning and land allocation; (c) forestry; (d) environmental management; and (e) financial accountability and trade investments.

14.2.1 Vietnam's Climate Change Strategies

Vietnam's National Target Program to Respond to Climate Change (NTP-RCC) focuses on accessing impacts on diverse sectors together with developing short-term and long-term plans. The target plan also signifies the need for sustainable development through low carbon economic trajectory. The Ministry of Natural Resources and Environment (MONRE) is responsible for the climate change program. The Ministry of Agriculture and Rural Development (MARD) has also launched sectorial action plans for climate change adaptation in agriculture and to enhance rural communities to benefit equally to adaptation and climate mitigation. Conversely, forest management for climate change adaptation is least recognized.

Nevertheless, directives have been issued under the national strategy on disaster prevention and reduction and execution of the Kyoto Protocol. Vietnam has also developed a national strategy on climate change to include the role of the forestry sector in climate change mitigation.

14.2.2 Land-Use Planning and Land Allocation

There are diverse interests over land and the need for legislative mandates for land use, ownership, and management are crucial to future REDD+ interventions.

- *Land-use planning*: Vietnamese laws on land-use planning are supportive of REDD+ implementation prominent through the order of land-use rights marking forest management. Locally, committees at each level commence land management in their dominions, prepare effective plans, and submit them to an upper hierarchy level for approval. Each ministry also develops their own sectorial plans, which are then combined with locally developed plans to develop a national land-use plan. The land is then correspondingly classified into diverse subgroups and forestlands. Specifically, forestland is divided into "production forest, protection forest, special-use forest, reforestation land, and land for afforestation." In terms of REDD+ implementation, production forest and land for afforestation may not be viable, but requires a naturally grown forest. Nevertheless, forestland has continuously been shifted for diverse purposes and subjected to illegal deforestation, creating risk for REDD+ permanence.

- *Mining*: According to IDLO (2011), "mining and mineral exploitation laws and regulations will be important for REDD+ as subsurface rights often take precedence over forest management and can cause reversals in the way in which forests are managed to reduce emissions." Because mining contributes 10% of GDP, Vietnam has established diverse funding incentives from the state budget (IDLO, 2011). A study conducted by Duc Quy (1996) shows that antimonite exploitation in Mau Due, gold and antimonite exploitation in Chiem Hoa, mining in Ban Lung, charcoal mining in Thai Nguyen, and stone extraction in Quy Chau have deforested approximately 25, 720, 218, 671, and 200 ha of forest, respectively. It is mandatory for mining and mineral extraction to conduct environmental impact assessments before implementation. The Law on Environmental Protection mandates rehabilitating affected areas primarily through replantations, together with depositing money for the degree of exploitation needed for rehabilitation. However, the actual implementation of this law is weak and deposits have not been actually collected (IDLO, 2011).

- *Agriculture*: Approximately 75% of people of labor age work in the field (General Statistics Office, 2011), as Vietnam is one of the biggest rice exporters in the world. Presently, a tendency toward plantation crops has converted natural forest into rubber and oil plantations; significantly posing encounters for REDD+. Though MARD has jurisdiction over both agriculture and forestry, it often produces policies that are not complementary. A good example could be the national strategy for rubber development that has a target of expanding rubber trees by 800,000 ha whereby some part of the natural forest will be cut down. This is coupled with the fact that plantations are likely to generate much more profit than through REDD+.

- *Protected areas*: Protected areas overlap with REDD+ activities. The Cancun Agreements demands REDD+ to consider multiple forest functions and conjoint ecosystem with conservation of biological diversity. But Vietnam is a signatory to various biodiversity conventions, resulting in 49% of Vietnam's threatened species being marked as "critically endangered" (World Bank, 2005). The Vietnamese government may propose a complete ban on forest extraction in protected areas likely to endanger the livelihoods of forest-dependent communities.

- *Land tenure and forest governance*: As mentioned earlier, the Law on Land (LL) and Law on Forest Protection and Development (LFPD) control the correlation between forest tenure and forest resource regulations, respectively. Recently, policies on forestland allocation promote both economy and conserving forests through development of local communities. According to LL, the forest belongs to the indigenous people while government is just the representative. However, LFPD gives to the state the right to manage and determine ownership of natural forests, which means the state has the authority to determine and exclude anybody for forest use. The state has granted forest through land allocations with due recognition of ethnicity and user's rights through forestland allocation programs. The LFPD confirms that forest owners are households, organizations, and individuals and the state releases forest mainly allocated to plant through recognizing plantation ownership and transferring forests from other owners. These owners could be involved in REDD+ benefits in the future. According to IDLO (2011), LL and LFPD offers rights through donation, lease, conversion, transfer, inheritance, and mortgage depending on the type of owner and type of forest. Communities only have use rights, they do not have full ownership rights, which may limit individuals' participation in REDD+ interventions.

14.2.3 Forestry

From 1943 to 1990 approximately 50% of Vietnamese forests were degraded. However, the total forest area has increased at 2.5% per annum from 2000 to 2005. Some scholars confirm Vietnam is going in the right direction, while others consider the increase to be a result of plantation forests that do not give attention to biodiversity. There are many forestry laws under the Vietnamese Constitution that confirm land and forest belongs to people and the government is just a representative.

The major framework for REDD+ implementation in Vietnam in the National Action Program on REDD+ 2011–12 that was approved by the prime minister in Jun. 2012 (Decision 799/QD-TTg). Other laws and policies that are relevant to REDD+ include (1) the LFPD, which comprises forest classification and rights of the government on protection and development; allocation forest to local communities, registering the forest use rights; property rights toward production forest and plantation forest; forest statistics and forest inventory; and forest protection responsibility; (2) Decision 178/2001/QD-TTg on the sharing of benefits between local household/communities participating in forest protection and management with different forest types: production, protection, and special-use forest; and (3) National Degree No. 99 and Decision 380 on Payment for Environmental Services (PES). The Law on forest protection and development-2004 (FAO, 2005) and other related policy documents have created favorable conditions for forest management, from centralized management regimes to participated forest management.

There were some new initiatives such as the support for forest plantation, payment for forest environmental services, testing the comanagement policies, and support for forest protection and management at the commune level. Those help to promote the economic development and effective usages of land and forestland. The rights of forest owners are expanded and ensured by the government, and have encouraged forest owners to feel safe in investing and developing production and trading to increase the effectiveness of forest and forestland usages. The governance, protection, and management of forest resources have been improved. The organizational structure was developed and strengthened. Forest planning and forest development strategies were implemented at the district and commune levels. There are several weakness of Vietnamese forest policy such as weak compliance and poor enforcement; unfair distribution and conflict (eg, SFEs vs. local people at the community and household levels; the private sector vs. local people); and the area of land allocated to households is small, of poor quality, and scattered.

14.2.4 Environmental Management

There are two significant laws in environmental management in relation to REDD+ implementation. These are the Law on Biodiversity (No. 20/2008/QH12), which states that proceeds from environmental services shall support biodiversity conservation and sustainable development (Article 74), and the Law on Environmental Protection (No. 55/2014/QHI3), which promotes sustainable uses of clean and renewable energy that reduces emissions of greenhouse gases (GHGs) and mitigates the impacts of climate change (Article 6).

14.3 KEY CHALLENGES TO REDD+ INTERVENTIONS IN VIETNAM

14.3.1 Clarity on These Policies

These national forest policies draw an overall picture of the forest management in Vietnam, including forest protection and conservation of biodiversity, forest development, forest extraction and utilization, forest investment, and credit and finance. It also stipulates the benefit-sharing mechanism and types of forest products that participants in forest management can benefit from, the PES activities, and the potential sources of sustainable forest finance the revenue from forest resources.

However, there are overlaps creating confusion among each other; for example, the Law on Environmental Protection (LEP), the Land Law (LL), and the Law on Biodiversity (LB). Although all these laws have as their objective the protection of forests, their use of inconsistent definitions of terms has led to complexity. These complexities are mostly seen in differences in land and forest classification and zoning and organization of special-use forests in these polices. REDD+ implementation based on these different legal backgrounds is likely to make the situation further chaotic due to different interpretations.

The Law on Forest Protection and Management (Article 4) defines and classifies forest into three types: protection, special use, and production forest. In each type, they are classified into different categories. This classification is complicated, creating overlap in identifying the usage purpose of each category, difficulty for forest protection, and management and utilization. For example, protection forest is divided into upper watershed protection forest, windbreak forest protection, flying san protection forest, dyke protection forest, and environmental protection forest. Within upper watershed protection forest, it is divided into very critical, critical, and less critical. (Decision 61/2005/QĐ-BNN dated 12/10/2005 of the MARD issuing criteria for protection forest classification.)

There are no regulations on timber and NTFPs harvesting in very critical, critical, and less critical protection forest. In addition, the classification of special-use forest in the Law on Forest Protection and Management (national park, natural reserve, species, and habitat protection area) are different from regulations on the classification of protected area in Article 16 of Law of Biodiversity 2008.

14.3.2 Extent to Which Law Recognizes Property Rights and the Right to Carbon

The state owns all forests in Vietnam, which are then allocated to households and organizations for short-term or long-term benefits. The LL also clarifies that land belongs to state ownership and may be granted for use. As a consequence, local stakeholders have no ownership and the only rights given are for use and management. This law is applicable to forest products and services.

However, the LEP clarifies the rights in other terms of forest protection that includes buying, selling, and transferring of GHGs shall be approved by the prime minister and this may challenge REDD+ design and implementations. The fact that current rights include carbon rights or not may pose additional challenge to REDD+; according to Covington et al. (2009), "uncertainty surrounding land title is significant preconditions for REDD+ scheme," To address this situation, legal certainty on land titles is essential. This may have consequences to ethnic minorities where no legal title is given and a huge gap is seen between de jure and de facto rights. In particular, legal and institutional challenges will only be ensured if the rights are managed equitably; otherwise, competing claims from the ethnic minorities are likely to arise and many of the communities may lose access to forest resources.

14.3.3 Extent of Conflict Resolution Mechanism

Conflicts are rampant in Vietnam, which sometimes result in physical damages. According to Sikor and To (2013), 70% of written conflicts are in regard to land disputes. The same source confers a statement by the Vietnamese National Assembly, "conflicts occur in many locations but are not adequately attended and resolved." The same study concludes the lack of an effective conflict resolution mechanism has failed to settle land conflicts locally. No policy talks about conflict resolution mechanism.

Most of the conflicts date back to the 1950s during SFEs and forest companies (FCs) restructuring to control land. Currently, even though FCs settle to handover land to locals, transfers are normally delayed because (a) financial and human resources lapses occur at local bodies, (b) most land is under private companies, and (c) land is too far away or not being productive. Generally, Land Use Right Certificates (LURCS) are supposed to solve conflicts, but they are futile as they were allotted devoid of attentiveness and their customary rights and established land users are neglected.

14.3.4 Extent to Which the Forest-Related Mandates of National Agencies Are Clear and Mutually Supportive

Vietnam's governance system is relevant to the country's legal preparedness for REDD+ because it determines the foundational mandates of the actors who design and implement REDD+ from the national to the local level. The government structure comprises the National Assembly (NA), the government, the people's courts, and the people's prosecutor. The state is managed at four levels: central, provincial, district, and communal. At the central level, the NA is the highest organ of state power, with the government as its executive body. Members of the government include 22 ministries and ministry-equivalent organizations.

At the provincial level, the PPC is a provincial body elected by and representing the local people. The People's Committee is elected by the PPC as an executive organ of provincial state administration. It is responsible for implementing the constitution, laws, and formal orders of state organs at the central level and the resolutions of the PPC within the province for the sake of socioeconomic development measures and national defence and security. Professional agencies, which are line departments of central level ministries or ministry-equivalent organizations, assist the People's Committee to realize its tasks.

At the district level, the District People's Committee (DPC) is the highest state body in the district and it has the task of law execution and is overall in charge of the forestland allocation process. The DPC is also assisted by professional agencies. At the communal level, Communal People's Committee is the elected body of Communal People's Council. The Communal People's Committee is in charge of executing constitutional laws, formal orders, and resolutions from the superior state body or communal People's Council.

However, there is a turf between MARD and MONRE in Vietnam. MARD is responsible for forest development and management while MONRE especially focuses on land and environment issues. Mostly, confusions and overlapping of rules often characterize differently in the implementation arena at both local, provincial, and district levels.

14.3.5 Extent to Which Polices Support Adaptive Forest Management

The initial forest policy clarified that 40% of total land area needs to be forested, while other related policies were formulated to support this policy. These policies were made by the government. Some of them are formed in the name of reducing natural catastrophes while others are simply by cabinet. The public is usually not provided with the details of the policies nor is any consent taken through the public. Most of the policies are ad hoc and there is no adaptive learning process involved. On the contrary, to improve forest conditions, REDD+ should be implemented in such a way that local needs and aspirations are not compromised. This fact cannot be envisioned through these policies.

14.3.6 Equity in the Distribution of Access to Forest Resources, Rights, and Rents

Forest land allocation: The LFPD (mentioned in Point b, Clause 3, Article 24—forest allocation) regulates that the government allocate natural production forest and get usage payment from all forest enterprises. However, most of the remaining natural forests are average and poor; there is almost no income from the forest within the long investment time frame (25–30 years). Resource-poor households dependent on these unproductive forests may face a hard time balancing their livelihood priorities with government usage payments.

Decision 304/2005/QĐ-TTg of the Prime Minister signifies "forest land allocation" provisions for local households and communities of ethnic minorities in the Central Highlands, and Circular 17/2006/TT-BNN of the MARD guides implementation of Decision 304, and clarifies the fact that usually unproductive forest low in timber stock and forest canopy is to be allocated (Sunderlin, 2006; Ngo and Webb, 2008).

Likewise, Degree 200/2004/NĐ-CP of the Prime Minister regulates that state-owned production could be taken back and further allocated to diverse stakeholders. Especially with the poor, household redundancy to invest in these forests may occur.

Forest lease: The LFPD (Article 25) has abided diverse payment time to different stakeholders, especially to the poor households that will be affected by these unequal arrangements. Though benefit sharing may be in place, there are serious possibilities for corruption, weak monitoring and enforcement and, most importantly, elites capture for productive forests (Balooni and Inoue, 2007).

Openness and competitiveness of procedures, such as auctions, for allocation of forest resources: Decision 178/2001/QĐ-TTg of the Prime Minister specifies the benefit-sharing mechanisms, rights, and responsibilities of individuals entitled to the government's land allocation and renting provisions with the following limitations:

- There were no clear criteria to distinguish between the products and by-products from forest.
- The ratio of benefit sharing (Articles 7 and 14) confirms that the allocation procedure is to be based on forest status at the time of allocation. Conversely, the policy does not mention anything about conducting forest inventory to determine the forest conditions both before and after allocations.
- In conditions whereby local authorities tend to conduct baseline inventories, the process is expensive and no financial support from the central state is rendered.
- No specific assignment of the bodies in charge of prior and post evaluation of forest resource for benefit sharing.
- Usually, the resource-poor are allocated forest in remote areas with low timber and NTFP stocks.
- No aforementioned time, place, price, or quantity exists in the distribution of benefits among the households.

14.3.7 Mechanisms for the Internalization of Social and Environmental Externalities From Forest Resource Use, Including Payments for Forest-Derived Environmental Services

There are several policy documents guidelines in Degree 99/2010/NĐ-CP of the Prime Minister on Payment for Forest Environmental Services, including:

- Circular of the MARD on the principle and methodology on determining the watershed area to serve for the payment of environmental services.
- Decision of the prime minister explicates on service, amount of payment, payment method, and benefit-sharing mechanism.

14.3.8 Transparency

Even though forest polices clearly address marketing and distribution methodologies and issues, market transparency is limited in the log trade. Small-scale producers often lack a clear understanding of the value of their timber crops and engage in limited negotiating. The government has developed a variety of ways to support local communities in managing their forest, albeit only in some places.

The present statistical reporting in forestry sector in Vietnam is above average; better than Thailand. Information is a powerful tool in managing the sector but was neglected in the past. The country has now developed some strategies for information management and much of the database on forestry sectors is made available to the public. However, for mid-level staff, they tend to perceive data collection as an administrative burden than a management tool.

However, rules on forest harvesting have not provided autonomy for production and marketing. The benefit-sharing mechanisms are still insufficient due to the fact that they have relied too much on the technical standards of forest management according to a centralized forestry model that has not accounted for the environmental service.

14.3.9 Forest Law Implementation

Usually financial and technical support from the government is lacking, which has compelled poor farmers to invest themselves in protecting and managing forests. The users are imposed a high tax and have no loan provisions coupled with strict harvesting and transportation rules. This is likely to reduce the willingness of local forest dwellers to participate and invest in natural forest management. Ironically, no government document has revealed improvements in the forest sector through the collected tax amount; usage of the forest resource tax is inconsistent. This may lead to the cynical situation that REDD+ will be successful in Vietnam.

14.3.9.1 Land Tenure Documentation and Administration

Law on Forest Protection and Management (Article 70) regulates the rights and duties of households, individual allocated natural production forest, as they can harvest NTFPs, have the rights to transfer, lease, mortgage, business capital contribution equal to the value added from allocated forest due to the investment of forest owners compared to the value of forest assessed at the time of allocation. Individuals can inherent their rights prescribed by law. Circular No. 38/2007/TT-BNN dated Apr. 25, 2007, of the Ministry of Agricultural and Rural Development instructing the process, procedure of forest allocation, forest lease for organization, households, individual and local communities; when doing the allocation of natural forest, it is required to determine forest stock. However, for the normal allocation process at the household level, the forest allocation area is determined along with generalizing the forest into type II, III, IV or rich, average, poor, restoration forest, there are no quantitative measurements on the amount and quality of timber on allocated forest. Thus, there are no baseline data to determine the added growth value of the forest. In reality, most ethnic minority households have yet to take full advantage of these policies. According to Tebtebba (2010), this is a difficult to solve social problem that poses challenges in the face of REDD+ implementation and the background that land allocation in Vietnam is based on the individual's financial capacity to invest. Especially with the ethnic minorities, the policy has often excluded them from receiving a larger share of land allocation (Linh, 2005).

14.4 LESSONS FROM EXISTING BENEFIT DISSEMINATION SYSTEMS

Equitable benefit distribution systems may be understood to initiate from the Cancun Agreements as a precursor to guidance and several safeguards. In particular, equitable benefit distribution aggravates benefits to directly involved communities from REDD+ contributing to overall forest management, poverty eradication, and sustainable development. It also delivers effective interventions and permanence by creating incentives for carbon sequestration through results-based payment. Vietnam does have experience in such results-based equitable benefit distribution systems through the 5 Million Hectares Rehabilitation Program (5MHRP) and PES programs.

14.4.1 Payment for Environmental Services

PES has been piloted in Lam Dong and Son La for the period of 2008–10 based on Decision 380/QĐ-TTg. In 2010, Degree 99/2010/NĐ-CP was released and become effective on Jan. 1, 2011. Up to now, 36 provinces out of the 63 in Vietnam have formed the Forest Protection and Development Fund (FPDF) and implemented the PES program (Vietnam Forest Protection and Development Fund, 2014). Under this program benefit users were required to contribute into a community fund which was then distributed for forest ecosystem management. As per IDLO (2011), "the level of payment for forest ecosystem services to households protecting the forest in the areas were three times higher than contracting fees of forest protection paid to State under 5MHRP." In response to the success of these projects, the government formulated Decree No. 99/2010/ND-CP on PES payment polices. Under this policy three funding types could be rendered for ecosystem services: (a) soil conservation and sediment retention, (b) conservation of water source, and (c) natural landscape protection and forest biodiversity conservation. There could be diverse funders: (a) hydropower facility, (b) pure water supply, and (c) tourism business organizations. These payment processes for ecosystem maintenance services are likely to establish an economic base to protect forest ecosystem and social benefits to communities.

Lessons from PES can aid in research and identifying issues on REDD+ implementation that significantly includes the degrees of compensation required to fully engage the community into PES schemes along with management and timely disbursement of funds; as such, this may serve for future REDD+ sharing mechanisms.

14.4.2 Five Million Hectare Reforestation Programs

Another example of a benefit-sharing mechanism may be built through 5MHRP. Basically 5MHRP primarily accelerated reforestation programs by planting and protecting new and existing forests together with promoting and protecting biological ecological diversity and foretelling a situation for sustainable development with the basic goal of achieving 40% forest cover rate. Additionally, the program sought to provide raw materials for forest industry development and create employment opportunities to contribute to Vietnam's poverty alleviation polices. In the process the government issued Decision 661QD/TTg to outline specific strategies to implement the program sustainably. These strategies comprised interventions for conserving natural forest; allocating forests to households, organizations, and individuals with sedentary cultivation and contributing to poverty reduction goals. The program has also envisioned goals, including

(a) To achieve regeneration of natural forest and plant new forests with sedentary cultivation to obtain 2 million ha of protected and special-use forest.
(b) To provide raw materials for forest-based industry and by newly planting 2 million ha of speciality crops and 1 million ha of fruit and perennial trees to obtain an overall 3 million ha of production forest.

Although 5MHRP achieved dual environmental and social benefits, future REDD+ implementation should account for some challenges posed by the program. First and foremost, although the government funded infinitely the fund was scattered. A clear monitoring process was missing, which would have been better through improved mechanisms for planning and auditing programs when needed. Last, but not least, the institutional structure did not provide an opportunity for incorporating local needs and aspirations; in a way, participatory decision making was missing. In the budget disbursement plans and costs for each component, fund grants were solely determined by the state. The real beneficiaries had the least possibility to decide the modes and degrees of fund disbursement for the program. For the future REDD+ mechanism, these challenges need to be preferably addressed.

14.5 CONCLUSION

This chapter provided a picture of Vietnam's current legal and institutional preparedness for REDD+. The chapter surveyed regulations, guidelines, and laws that were thought to be significant to the study and found that there are some institutional and legal challenges for REDD+ intervention; simultaneously, it featured some innovations like PES and 5MHRP too. Vietnam is well placed to extract benefits from REDD+ for its strong commitments in achieving goals and promoting national developments. The main drivers for deforestation are conversion of paddy fields to plantation crops, lack of clear forest rights, and increased demand for forest products. Owing to the fact that forest cover resulted in an increment to 38.7% shows Vietnam's capacity and determination in sustainability of forest and improving livelihood simultaneously.

REDD+ is situated within a general governance framework that includes laws for local participation and access rights to information and projects possibilities for informed policy and decision making. Vietnam has also achieved success through its *doi moi* economy, but contradictorily achieved this through intense natural forest exploitation. A range of national institutions and laws are pertinent to REDD+ interventions and development, including land-use planning, forest management, and so forth. However, confusion about existing laws have left 100 bylaws contradictory and overlapping. This may create huge problem for REDD+ design and interventions as diverse stakeholders may interpret it differently, further deteriorating the forest. The fact that there is no special policy in Vietnam distinguishing the rights of indigenous peoples, who live largely in the mountainous region ranging from the north to the south, may hinder the real needs of the people. In particular, legal and institutional challenges will only be ensured if the rights are managed equitably; otherwise, competing claims from the ethnic minorities are likely to arise and many of the communities may lose access to forest resources. Forest conflicts are high and could undermine successful REDD+ mechanisms that affect the sale and increasing carbon stocks and overall ensuring legalized timber harvesting. Likewise, lack of capacity to undergo sustainable forest management and subsequent livelihood needs may lead to further conversion of agricultural land to rubber plantations, which usually does not increase carbon stocks. The failure to identify customary rights and enhance participatory decision making may lead to noncompliance with the international REDD+ arena.

It is crucial to consider that, currently, Vietnam needs to undergo various institutional and policy changes backed up by more legal research and assessment. REDD+ may offer opportunities to overcome many of the issues if ways are found to identify local perceptions and needs. As noble approaches in policies are recently emerging and new answers are freshly tested, a great technical assistance will be required to provide tailored solutions that are specific and appropriate. It is equally important to engage local livelihood issues. Lastly, Vietnam is legally set for REDD+ but requires further attention to legal and regulatory measures, which is possible through addressing the challenges mentioned above. Legal and regulatory improvements require support for cooperative solutions typically from the ground up.

References

Balooni, K., Inoue, M., 2007. Decentralized forest management in south and Southeast Asia. J. Forestry 105 (8), 414–420.

Bojo, T., 2011. Vietnam Development Report 2011: Natural Resources Management. World Bank, Washington, DC. Retrieved from, http://documents.worldbank.org/curated/en/2011/01/15768936/vietnam-development-report-2011-natural-resources-management.

Covington, Baker & McKenzie, 2009. Background Analysis of REDD Regulatory Frameworks. The Terrestrial Carbon Group. The Terrestrial Carbon Group and UN-REDD Program. Retrieved from, http://theredddesk.org/sites/default/files/resources/pdf/2010/Background_Analysis_of_REDD_Regulatory_Frameworks.pdf.

Duc Quy, N., 1996. Deforestation in Some mining Areas. Science, 8.

FAO, 2005. Vietnam: Law on Forest Protection and Development- 2004. FAOLEX-legislative database of FAO legal Office. Food and Agriculture Organization of the United Nations. Retrieved from http://faolex.fao.org/docs/pdf/vie50759.pdf

Food and Agriculture Organization of the United Nations (FAO), 2010. Global Forest Resources Assessment 2010: Country Report of Vietnam. Retrieved from, http://www.fao.org/docrep/013/al664E/al664e.pdf.

General Statistics Office, 2011. Employed Population at 15 Years of Age and Above as of Annual 1 July by Residence. Statistical Documentation and Service Centre—General Statistics Office of Vietnam. Retrieved from, www.gso.gov.vn.

IDLO, 2011. Legal Preparedness for REDD+ in Vietnam: Country Study. International Development Law Organization (IDLO), Food and Agriculture Organization of the United Nations (FAO) and the UN-REDD Programme.

Law on Land, 2013. Vietnam Land Law 2013. The National Assembly: Socialist Republic of Vietnam. Vietnam Law in English. Retrieved from, http://vietnamlawenglish.blogspot.com/2013/11/vietnam-land-law-2013-law-no-452013qh13.html.

Law on Protection of the Environment, 2001. Law on Protection of the Environment. Socialist Republic of Vietnam. Retrieved from, http://haiduong.eregulations.org/media/Law%20on%20Enviroment.pdf.

Linh, D.T., 2005. Forestry, Poverty Reduction and Rural Livelihoods in Vietnam. Ministry of Agriculture and Rural Development, Hanoi, Vietnam.

Ngo, T.D., Webb, E.L., 2008. Incentives of the forest land allocation process: implications for forest management in Nam Dong district, Central Vietnam. In: Webb, E.L., Shivakoti, G.P. (Eds.), Decentralization, Forests and Rural Communities: Policy Outcomes in South and Southeast Asia. Sage Publications, New Delhi/Thousand Oaks/London.

Nguyen, H.H., 2013. Transition to sustainable forest management and rehabilitation in Vietnam. In: Paper presented at International Symposium on Transition to Sustainable Forest Management and Rehabilitation: The Enabling Environment and Roadmap. 23–25th October, Beijing, China.

National report at the United Nations Conference on Sustainable Development (RIO+ 20), 2002. Implementation of Sustainable Development in Vietnam. Socialist Republic of Vietnam. Retrieved from, https://sustainabledevelopment.un.org/content/documents/995vietnam.pdf

Sikor, T., To, P.X., 2013. Conflicts in Vietnam's forest areas: implications for FIEGT and REDD+. Retrieved from, http://www.forest-trends.org/documents/files/doc_4211.pdf.

Sunderlin, W.D., 2006. Poverty alleviation through community forestry in Cambodia, Laos, and Vietnam: an assessment of the potential. Forest Policy Econ. 8, 386–396.

Tebtebba, 2010. Indigenous People, Forest and REDD Plus. State of Forests, Policy Environment and ways Forward. Tebtebba, Philippines. Retrieved from, http://redd.unfccc.int/uploads/63_13_redd_20110523_tebtebba_state-of-forests.pdf.

Vietnam Forest Protection and Development Fund, 2014. Report on the implementation of PES program following the degree 99/2010/NĐ-CP dated 24/9/2010 of the government.

World Bank, 2005. Vietnam Environment Monitor 2005: Biodiversity. World Bank, Washington, DC. http://documents.worldbank.org/curated/en/2005/12/6451376/vietnamenvironmentmonitor-2005-biodiversity.

Social and Gender Issues in Land Access and Vulnerability in Ky Nam Commune

H.T.A. Phuong*, T.N. Thang†

*Hue University of Sciences (HUS), Hue, Vietnam †Hue University of Agriculture and Forestry, Hue, Vietnam

15.1 INTRODUCTION

15.1.1 Background and Purpose

Land always has been a fundamental and important asset for rural people in developing countries, and therefore secured access to land can effectively improve and enhance these people's socioeconomic status and well being (De Soto, 2000; Do and Iyer, 2008). In Vietnam, where about 70% of the national population is residing in rural areas, land-related production activities have traditionally played a major role in livelihoods, as main sources of home food consumption and for income. Therefore, access to land is absolutely vital for these people. This is even more critical for women because, compared with men in rural areas, women's livelihoods and contributions to households' welfare traditionally have been dependent on their working on the land, as widely argued (FAO, 2010; UNDP, 2013; Menon et al., 2013).

Realizing the importance of rights to property or resources in governing natural resource management as well as in improving the welfare of individuals and households depending on natural resources for livings, since the early 1990s the Vietnamese government has issued a number of land-related policies and laws enabling the development of rural livelihoods and the rural economy. In particular, for both arable and forestland individual households on the ground have been allocated a certain piece of land for their productive purposes on a long-term basis of use, and land-use certificates have been granted to these landholders. As a result, the access to land of rural people has been enhanced and strengthened more than ever as UNDP (2013) stated that about 90% of rural households in Vietnam got land and land-right certificates that could guarantee their secured access to land. Thanks to such access to land through granted legal rights to resources, rural livelihoods are secured, food consumption is provided, and income is improved. Then, rural poverty is enormously reduced and socio-economic conditions of rural areas and the whole nation are enhanced as widely reported in many related reports and studies (eg, Ravallion and Walle, 2003; FAO, 2010; Kirk and Nguyen, 2009; Dang et al., 2006).

However, it has been widely accepted among scholars and researchers that land access is not necessarily a smooth process and access to resources, particularly land, are socially and gender differentiated and very context-specific (eg, Von Benda-Beckmann et al., 1996; Cooke, 2004; FAO, 2010). Built on such ideas, this study through a gender lens will focus on investigating social and gender differentiation in land access among rural households and its implications on vulnerability differentiation of these households in practice. Specifically, through a case study in a rural community in Central Vietnam, this study examines the differences and changes in practical holdings to arable land and forestland of particular social groups of households and the underlying causes of such differences and changes. In addition, it analyzes how such situations of access to forestland and arable land of these groups of households both creates and marks vulnerability differentiation in terms of livelihoods and income among them. It may be argued that a rights-based property regimes system is not adequate enough to guarantee secured access to land for local people and that gender-specific characteristics are the markers of differentiation in land access among local households. The ultimate purpose of this study is to highlight social and gender dynamics in natural resources access and the mismatch between policy and practice in natural resources distribution and management.

The sections below will discuss a brief analytical framework, to be followed by a discussion on the study site and research methods. Empirical sections will present and discuss the processes and current status of land access of particular social groups of households, the underlying causes, and their implications on marking vulnerability to livelihoods and incomes of these households, which will be followed by the discussion and conclusion section.

15.1.2 Access to Resources and Vulnerability: Property Rights Regime, Power, and Gender

Although the concept of access often has been used in property theories and analysis, it has always been under debate among scientists. In fact, there have been two dominating points of view in theorizing access. The first one is defining access as a *right* to enter a defined physical property or resource and obtain benefits from such property (eg, Schlegar and Ostrom, 1992). Such a right refers to an "enforceable claim" that is socially acknowledged through law, custom, or convention. The second point of view considers access as the *ability* to benefit from things (eg, Ribot and Peluso, 2003). Ribot and Peluso (2003) stressed that ability is related to power, which is embedded in social relationships reflecting the capacity of some actors to affect the practices and ideas of others and containing socially and legally forbidden acts. In other words, the issue of rights to resources is a matter of law, whereas the issue of ability goes beyond it and reflects social elements as mediators of resource access rather than legal rights only.

Ribot and Peluso (2003) stated that there are different types of mechanisms regulating the access to resources of people ranging from legal to social domains. Within the legal domain, a person can access resources through being granted with rights by the state or central governments and manifest by legal titles or legitimatized claims over certain resources. It is widely agreed that with such a rights-based mechanism, access to resources will be more secure and the holder will have more incentives to invest in that resource in the long term (Hoang, 2009). Within the nonlegal or social domain, the ability of a person to access resources is regulated by a diverse set of elements including labor, income, credits, or equipment that can be used for investing in extraction, production, or labor mobilization for gaining benefit from resources. For example, those who have access to labor can gain benefit from a resource or an opportunity at any stage where labor is required. Meinzen-Dick et al. (1997) also argued that property rights analysis has often focused on rights granted and held by a person or a household without a recognization of the differentiation between property holders and factors mediating such differentation, including gender, age, and other characteristics. Naidu (2011) argued that access to resources is not social or gender neutral irrespective of equal rights that individuals and households claim. Through some research findings, Sikor and Lund (2009) further argued that even formal property rights do not necessarily ensure the access of farmers who hold rights to derive material benefits from such natural resources through showing a case on poor and nonpoor households in the context of forest devolution in Vietnam (see more in Sikor and Nguyen, 2007).

Irrespective of legal or social mechanism, there have been many discussions about the existence of power as a key determinant in land access of people. In the formal relationship domain, power over resources reflects the power of authority. In fact, the government still remains the ultimate power holder of resources. Vietnam is a typical case. The fact that all natural resources in Vietnam are defined under the possession of the whole nation and the central government is the powerful legal representative is written in all laws and legitimized documents related to natural resources. Specifically, as indicated in the Vietnamese land laws (particularly Land Law 1993), land is considered as the possession of the whole population and the government has full rights to this asset in terms of (i) deciding the purpose of land use; (ii) granting land to users; (iii) withdrawing land; (iv) converting land-use purposes; and (v) setting land price. At the operational level, local farmers are granted use rights to land for productive purpose for 20 years for arable land and 50 years for forestland with an issued "Certificate of Land Use Right" (Red Book). As indicated in Land Law 1993 and 2003, farmers can obtain five rights to their "own" land including (i) the right to exchange land, (ii) the right to assign land, (iii) the right to bequeath, (iv) the right to lease land, and (v) the right to mortgage and guarantee. By these rights, it can be said that landholders on the ground have a "bundle of rights" to land. However, central governments still have full power in distribution and management of natural resources regardless of it granting a "bundle of rights" to local people.

In other domains outside legal institutions, power reflects the asymmetries in access to other resources and social relationships among people for getting benefits from a resource. In this domain, gender is regarded as a key dimension, which on one hand reflects the differences in practice in itself and on the other hand the outcome of the practice. For instance, a case study in Agrawal (2007) highlighted the gender differences in land access even if property rights exist in practice.

During its development process, gender has been always conceptualized as a homogenous term equivalent to either "women" or "men and women" as homogeneous groups. Cornwall (2007) argued that gender always has been used as a descriptive term that tries to enumerate the differences in division of labor, in power distribution between men and women, and to argue gender issues in practice rather than as an analytical term that looks at the causal structure shaping gender relations and identities. Taking the advances from the existing points of views on gender, this study first also views gender as a type of social difference between human beings. It then extends the concept of gender beyond the differences between men and women as often assumed, and it fundamentally understands gender as a key variable contributing to shape particular women or men with particular social status in society. In other words, gender is only one among axes of social differences and regulated by these different socioeconomic characteristics. More specifically, it focuses on investigating socioeconomic differences in terms of class and headship that shape or stratify individual women and men.

These above-discussed dimensions of access frame this study. First, we concur that access to land is a combination of both acknowledged rights and the ability of particular individuals to obtain land for their productive purposes. And finally, we consider access to land as a social and gender differentiated process and largely shaped by legal and socioeconomic conditions and processes, which may in turn have unequal outcomes for those related.

15.2 METHODOLOGY

15.2.1 Study Site

This study is part of the research for the author's doctoral study. Ky Nam Commune of Ky Anh District in Central Vietnam was selected to conduct field-based activities from Jun. 2010 to Oct. 2011. Located in the far southeastern part of the district and the province (see Fig. 15.1), this commune is surrounded by the sea in the east and Hoanh Son Mountains of the Truong Son mountain range running along the border with Laos in the west, south, and North, spanning an area of 1801 ha. This characterizes the commune as both coastal and mountainous.

With such a distinct physical location, Ky Nam is susceptible to storms, droughts, and localized southwestern feohn (chinook) winds in the dry season (Jan. to Aug.), and heavy rains and floods during the annual rainy season (IPONRE, 2009). Due to its distinct physical location, such climate events are more extremely severe in this site than elsewhere within the region. Particularly in the dry season, the deficit of surface water supply for use and production has been a recurrent phenomenon for generations. According to current scenarios on climate change and sea level rise for Vietnam, this commune will be among the most vulnerable regions to increasingly high temperature and dwindling rainfall as well as sea level rise (MONRE, 2011).

Since the early 1990s, along with socioeconomic transformations nationwide as resulted from the *doi moi* revolution, this locality has been experiencing institutional and socioeconomic changes related to resource distribution and use. Particularly in the early 1990s, arable land was allocated to individual households instead of cooperatives as usual for productive purposes for 20 years running. In early 2000, the authorities from Ky Anh District confiscated nearly 100 ha, of which nearly 80% was cropland being cultivated by local households for establishing shrimp farming in this commune. So far, this production activity has been under the access and control of nonresidential people. In addition, the policies on forest protection and forest allocation have changed the practice of forest use and management. Natural forest is no longer open access land, whereas barren forestland has been allocated to individual households to cultivate forest plantations. These structural changes have made residents all the more vulnerable to those who impinge on their livelihoods, which will be discussed in the section on results.

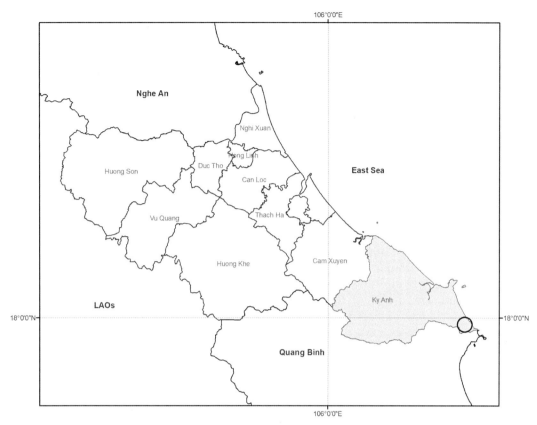

FIG. 15.1   Location of the study site.

15.2.2 Methods

A mixed methods approach, combining both qualitative and quantitative data and analysis, was applied in this study. Qualitative data was collected from group discussions, key informant interviews, in-depth interviews, and observation; whereas, a household survey generated quantitative and stratified information.

Separate focus group discussions were held with two groups of adult people including representatives from different headship- and class-categorized households for their differentiated perception on the importance of land for their livelihoods,[1] and their possible advantages or disadvantages in accessing land. The interviews with key informants such as the chair of the Commune People's Committee (CPC) and elderly persons shed light on past and current practices of resource use and management, particularly land resource, and local resource-related socioeconomic and institutional changes. In-depth interviews were conducted with some households from each household headship and class categories to shed light on

[1] In this study, livelihood refers to activities that provide an income both in-cash or in-kind.

TABLE 15.1   Total Population and Sample Size for Ky Nam Commune for Household Survey
by the Year of 2011

	Total Households		FHHs		MHHs	
	Total Population	Actual Sample	Total Population	Actual Sample	Total Population	Actual Sample
Whole commune	634	280 (44%)	152	58 (38%)	482	222 (46%)
Poor	242	109 (45%)	84	32 (38%)	158	77 (49%)
Nonpoor	392	171 (44%)	68	26 (38%)	324	145 (45%)

Ky Nam's Communal People Committee. Annual Socio-economic Reports from 2000–2011. Vietnamese version.

the dynamics of land access, underlying causes marking their differentiations in land access, and the possible vulnerability they have when they are facing obstacles in land access for developing livelihoods.

The study conducted a household survey with 280/634 households of the commune to gather information from a broader sample size on household resources, patterns of land holdings, and factors affecting their access to land among local households. For the sampling procedure, we based it on the available list of households[2] provided by the Commune People's Council, first of all stratified local households utilizing class and household headship categories for the purpose of a gender and class intersectionality analysis. Then we conducted a random sampling selection using RAND() in Excel in each category of households and, as a result, we arrived at the sample size (see Table 15.1 for the sample size in each substratum).

The data from the household survey is input and analyzed in the Statistical Package for the Social Sciences (SPSS) software, which is presented in diagram, tabulation, and so on. For the data from qualitative methods, it mostly used the content analysis to yield the results based on the objectives of the study.

15.3 RESULTS

15.3.1 Some Major Institutional Changes Related to Land Distribution and Use in the Research Site Over Time

Although this commune was established in the late 1970s,[3] it was not until the early 1990s, thanks to the implementation of Land Law 1993 nationwide, that local households in Ky Nam Commune were allocated with cropland[4] and at the same time granted legitimatized

---

[2] In Vietnam, households are annually listed with brief information on household headship and class for administrative management. At the time of survey, the commune had a total of 634 households.
[3] Before the 1990s, arable land was under the control and management of the cooperatives. Local people worked on land as members of those cooperatives and got some benefits based their labor contribution.
[4] Per the regulation, each household would be allocated a certain parcel of cropland according to their household size (each household member got on average 330 m^2).

Land-User Right Certificates (Red Book). Such arable land really made sense to local households because this was the first time they could access land for their own productive purposes rather than working on land for the cooperatives to get the same returns as before. However, as was assumed previously, land access is not necessarily a smooth process, and the practice of land access among social groups of households that results from this process can be complicated, which will be discussed in detail in the next section.

Another change related to land resources in this locality was the establishment of a shrimp farming project[5] in 2001. In fact, this project was initiated by Ky Anh District People's Councils, which took nearly 100 ha of local land, of which nearly 80% was cultivated land and most of it was high-yield riceland at the time.[6] The data in Table 15.1 displays an increase of land for shrimp farming to 96 ha and a decrease of cropland from 96 ha down to only 32 ha after 2001.

In terms of forestland, since the early 2000s the policies on natural forest protection and forestland allocation have been implemented in this site as an effort of the central government to protect natural forest and increase forest cover. As a result, local people have been prohibited to access natural forest for any purposes as they used to do before. In addition, the allocation of forestland contributed to increase the area of owned allocated forestland among local households from only 2 ha in early 2000 to 239 ha in 2009 (see Table 15.2).

In short, Table 15.2 shows the changes in land-use pattern in this commune over time as a direct result from the visible socioeconomic changes just discussed. Particularly, the size of some types of land use such as riceland and natural forest decreased whereas other types of land use such as other annual cropland, forestland, and land for aquaculture production significantly increased. The following section will discuss in detail how such institutional changes affect land access among different groups of households with particular focus on the comparison between female-headed households (FHHs) and male-headed households (MHHs) of different class categories.

---

[5] This productive sector is owned and explored by outside people and currently a private company instead of the head of the local population. In reality, the managers of this shrimp farming project promised to use local laborers in the shirmp production; however, when the project was operated, very few local people were hired to work because they were not qualified enough.

[6] As laws, the central government has full rights on natural resources; particularly, they have rights to withdraw used land for other purposes. In addition, by the early 2000s the people's councils at the district and provincial levels were granted with more rightful power in terms of resource use and management through the process of rights decentralization (Decree 66/2001/NĐ-CP on Sep. 28, 2001). At that time a national development strategy on the conversion of low-productivity land to high-quality activities, particularly aquaculture, had been popularized nationwide. The people's council at the district level with their granted power successfully initiated the project on shrimp farming in Ky Nam Commune based on the claim that arable land in Ky Nam Commune yielded poor productivity.

The sale of arable land for shrimp farming activities was conducted through legal contracts with individual households who have land area within the shrimp zone. Land was measured and categorized specifically for individual households. The price was based on the state price instead of the market price (the state will announce a monthly or yearly price for different types of goods or services, which is always lower than that of the free market). Individual households got annual payments in cash for a specific land area until their user rights to land expired (in the year of 2014).

TABLE 15.2   Pattern and Changes in Land Use in Ky Nam Commune by 1990, 2000, and 2009

Land Type	1990		2000		2009	
	Hectare	% of total	Hectare	% of total	Hectare	% of total
Total natural area	1842	100%	1842	100%	1842	100%
Riceland	151	8.2%	94	5.1%	32	1.7%
Other annual crop land	–	–	50	2.7%	60	3.3%
Perennial land	8.13	0.4%	27	1.5%	48	2.6%
Forest	648	35.2%	431	23.4%	376	20.4%
Forest plantation	–	–	2	0.1%	239	13.0%
Aquaculture	–	–	–	–	96	5.2%
Residential land	56.26	3.1%	252	13.7%	294	16.0%
Unused land	978.61	53.1%	986	53.5%	697	37.8%

Ky Nam's Communal People Committee. Annual Socio-economic Reports from 2000-2011. Vietnamese version.

15.3.2 Access to Land for Livelihoods: Where Power and Gender Matter

15.3.2.1 Land Allocation Process and Unequal Access to Arable Land Among Households

Thanks to the land allocation process since the early 1990s, local households could obtain their "own" land for productive purposes. As regulated, each member of a household could get 330m^2 of riceland, meaning that the size of arable land a household could get depended on the size of that household. In addition, local households could access land for productive purposes through other means including reclaiming[7] unused land and inheritance.[8] The statistical data from the household survey presented that a sampled household got averagely 5.6 sao[9] of arable land from different means of access, of which 3.9 sao is for rice cultivation and the remainder of 1.7 sao is peanut or cassava cultivation (see Fig. 15.2).

Fig. 15.2 also showed the differentiation in land access among different types of household. On average a MHH and a nonpoor household obtain a larger size of cropland as compared with a FHH and a poor household, respectively. Within each headship, the size of cropland of the poor is also smaller than that of the nonpoor. Both data from qualitative and quantitative methods reveal some accounts explaining the differences in terms of cropland holdings between MHHs and FHHs. First, the household size of a MHH is normally larger than that of a FHH; therefore, they can access a larger size of cropland. Secondly, with at least one actively economic male laborer as the household head, a MHH could expand their cropland by reclaiming unused land. Finally, consistent with Vietnamese traditional culture in terms of inheritance, a man is much more likely to get some cropland from his parents than a woman when married. This is consistent with the statistical results from the household survey that 20.7% of MHHs and 13.3% of FHHs got cropland through inheritance from their parents and relatives.

---

[7] Reclaimed unused land is also recognized by local government.
[8] This right of land access is recognized in land-user rights granted by the central government.
[9] 1 sao = 500m^2.

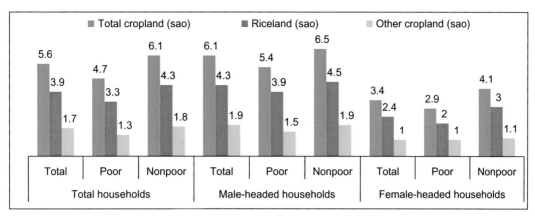

FIG. 15.2 Mean values of land holdings of sampled households by gender and class that resulted from the land allocation process. *Household survey, 2011.*

Within class-categorized households, the differences in cropland holdings among households are mainly caused by the intrinsic characteristics of households. The study found that poor households, regardless of being male-headed or female-headed, tend to have at least an unhealthy or disabled member or few in the labor force to work on land, which plays as influencing role in constraining these households in accessing land through both legal mechanisms and reclaiming activity. Mrs. Phin, the female head of a poor household, for instance, is now 53 years old and a single mother with a child. When she was young, she became disabled from a war bomb in 1968. She confirmed that her poor health and no labor force experience at that time denied her from being allocated with land. She said "Local authorities said that I was not capable to work on land, therefore they did not allocate land to me."[10]

In short, the land allocation process caused inability as well as inequality in access to cropland, especially riceland, among households in this local area. The MHHs of class category with their gendered advantages in terms of actively economic male labor, the ability to reclaiming unused land, and to access inherited land have larger holdings of arable land as compared with the FHHs.

15.3.2.2 The Worsening Impacts of a Shrimp Farming Project on Cropland Holdings of Local Households

As previously mentioned, the establishment of a shrimp farming project in 2001 took out a large part of riceland area from this locality. Seventy percent of sampled households involved in the household survey confirmed their land loss for shrimp farming, on average 3.4 sao (the smallest 0.2 sao and the largest 15.5 sao per household).

In reality, whether a household lost their cropland or the size of lost land is basically due to the location of their cropland plots rather than being caused by any other factors. With a larger size of cropland, a MHH and nonpoor household having cropland lost a larger size of

[10]In-depth interview with Mrs. Phinh, a poor female head of household in Ky Nam Commune, August 6, 2011.

cropland as compared with a FHH having cropland. Some households even lost nearly all their cropland, such as the case of Mr. Que. After the land allocation process, his family had a total of 6 sao of cropland. However, he lost 5/6 sao of cropland for shrimp farming activities and the 1 sao left is used for rice cultivation to provide a part of the household's rice home consumption.

FHHs and poor households, despite their loss of smaller sizes of cropland, originally had smaller sizes of cropland than the MHHs and the nonpoor. Their landless position even worsens their situation of land holdings as showed in Fig. 15.3.

15.3.2.3 *Forest Policies: Alienation of Traditional Forest Access and Unequal Access to Forestland*

Forest traditionally served as a place for local people, especially women, to collect fuel for home use and for sale to get petty income and as a place for sheltering and grazing their animals in severely cold and drought times. However, starting in early 2000 and especially since the enforcement of the governmental law No 29/2004/QH11 regarding forest preservation and development that legally prohibits the local population to access forest for any purposes. As a result, local people's access to natural forest was suddenly stopped, and their source of income was lost accordingly. People now either had to look for fuel wood from other sources such as their garden trees or even buy from the market or access natural forest illegally for these products. In other words, the local governments have withdrawn the traditional rights to forest access by the local population.

At the same time, the government began allocating barren forestland to individual households. Local households who had demand for forest plantation would be allocated with some area of land as requested. In 2009, 239 ha of forestland was recorded to be allocated to hundreds of individual households in the locality (see Table 15.3).

At the time of the study, the results from the survey displayed only 44% of the total 280 sampled households were forestland beneficiaries, among which 49% came from MHHs and

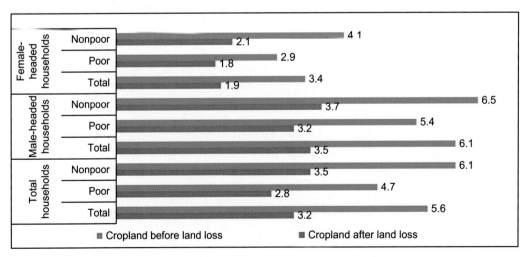

FIG. 15.3   A comparison of cropland holdings of local households before and after land loss by headship and class. *Household survey, 2011.*

TABLE 15.3 Forestland Beneficiaries and Average Forestland Holdings (Sao/household) Among Sampled Households by Headship and Class

	Total Households			FHHs			MHHs		
	Total (n = 280)	Poor (45%)	Nonpoor (44%)	Total (n = 58)	Poor (38%)	Nonpoor (38%)	Total (n = 222)	Poor (49%)	Nonpoor (45%)
Forestland beneficiaries	44%	43%	45%	26%	28%	23%	49%	49%	49%
Average forestland size	23.1 (1–260)	14.3 (1–140)	28.5 (1–260)	11.9 (1.5–60)	17.1 (2–60)	4.1 (1.5–10)	24.7 (1–260)	13.6 (1–140)	30.6 (1–260)

Household survey in Ky Nam, 2011.

only 26% from households with female heads (Table 15.3). The forestland size (ranging from 1 to 260 sao) was also noticeably variable among households based on headship, in which the average forestland size for a household with a male head was nearly twice as large as a household with a female head. Mrs. Truong[11], a female head of household, affirmed that she stayed in the local area with her young son and engaged in crop cultivation and petty trading that brought her a rather good income, whereas her eldest daughter has done permanent out-migration to the south for a job. When forestland allocation was available, she also applied for only one sao of forestland but just planted some trees there for uncommercial purposes and currently she lets her brother work on that forestland. Qualitative data from some other female heads also showed their similar reasons for applying for a very small size of forestland. In contrast with female heads, MHHs tend to apply for larger sizes of forestland for forest plantation to increase their income, as in the case of Mr. Phia, a male head of household. With three active laborers, his family got 40 sao of forestland since the early 2000s and invested in doing forest plantations. At the year of study they got 30 million VND after their second harvest.

Both quantitative and qualitative data showed that the availability of labor (especially male labor) and financial capacity reflected by annual household income (excluding forest plantation income) play their importance in the decision of getting forestland as well as the size of forestland of a household in this commune. In reality, as compared with male heads, female heads had on average a smaller labor size (especially due to the absence of an economically active man) and lower financial capacity; thus, the chances of being allocated forest land and the size of allocated forest land—if at all—were therefore all remarkably smaller.

In other words, forestland allocation policies caused unequal and uneven access to forestland between households particularly between MHHs and FHHs, which could in turn affect inequality in gaining entitlements from forestland among these households.

These findings show that local households are suffering from a very insecure situation of land access. In addition, the differences among different groups of local households in land access are evident, for different reasons.

---

[11]In-depth interview on Aug. 10, 2011.

15.3.3 Vulnerability Differentiation of Local Households to Insecurity of Land Access

Due to the importance of land-based livelihoods, particularly rice production, for households' well being. the insecurity in land access along with the institutional changes in the research site since the early 1990s has had adverse implications on the livelihoods of local people. If many local households got land for their own use after the land allocation process, the establishment of shrimp farming in 2001 makes these households either decline or lose their land access. The household survey data shows that nearly 10% of sampled households became landless after the land loss for shrimp farming after 2001. It means that these households lost a traditional means of rice supply and income, as one woman said.

> I relied too much on crop production for our living. Although I had a small size of rice land but it was good land and brought me high yield. With agricultural by-products, I could raise some pigs for extra income. Unfortunately I lost totally this small land because of the shrimp farming and suddenly became landless and unemployed in agriculture.[12]

Nearly all landless households affirmed that the loss of all riceland for shrimp farming also forced them to stop livestock rearing, as the predicament of the women below shows:

> Loss of land for shrimp farming put us in a very awful situation. We suddenly could not work on land as we did for generations, could not produce rice for feeding our families and even had to sell out all pigs and poultry because of we had no feed stuff.[13]

Other households, although they did not lose all of their land to shrimp farming, felt it was very difficult to continue their lives with their traditional livelihoods. Apart from the landless people who totally lost their livelihood, the livelihoods of those who lost a large part of riceland also became more insecure, especially in terms of its function of providing a rice supply for home use. This is because most of the lost land was irrigated riceland, which was producing two crops per year and generated good yield, more than the rainfed riceland. Due to the loss of this large part of irrigated riceland, local households stopped growing the second crop starting from the year 2002, and the cultivation of the first crop was on a smaller land area than before. Qualitative data from the field indicated that the cessation of cultivating a second crop resulted partly from land loss to shrimp farming and partly from climate-related water scarcity increasingly occurring in this local. Therefore, land-based livelihoods also became more insecure than ever, even for those who were not landless.

The loss of traditional access to natural forest as a result of forest protection policies suddenly led to serious inconvenience for local households' lives, as the female group expressed.

> Since early 2000 we have not been allowed to access forest as before. This really puts us in very adverse condition because we suddenly lose our traditional access to forest.... No fuel wood for home use, no extra income for food, no grazing place for animals....[14]

---

[12] Mrs. No, a poor female head of household in an in-depth interview, Jul. 28, 2011.
[13] Mrs. Thuy and Mrs. Sanh, in a group discussion with female farmers, Jun. 21, 2011.
[14] Mrs. Khoa in group discussion with female farmers on Jun. 21, 2011.

The households who often had holdings of more riceland also seemed to have the biggest loss, as in the case of Mr. Khan's family[15] who lost 91% of his total irrigated rice (15.5/17 sao) to the shrimp farming project. In the past, rice production not only ensured his family's rice demand but also gave a surplus for sale as their main source of both in-cash and in-kind income. All his family's labor force used to work only on the land. But now, left with only a small plot of land and only one rice crop per year, these family members suddenly lost their jobs in agriculture, especially during the lean time after the first crop.

Decreased land size also brought fewer agricultural by-products than before, and resultantly affected the raising of pigs and poultry, as the predicament of some women below shows:

> We had to reduce raising pigs for extra income because rice bran—the main feed source for feeding pigs—was not as plentiful as before due to the small size of riceland. Poultry also faces the same fate. Many women in this commune did like us. We had not imagined this awful situation when land has become so scarce.[16]

As a result of a decrease in land size, local people also faced another problem of the decrease in rice productivity. The recorded data from the CPC indicated that as direct consequences of land loss for shrimp farming local rice production dramatically dropped from 240 ton in 2000 to 39 ton in 2002, extremely threatening the rice supply for the local population. When surveyed, many households expressed their sadness with the fact that their rice productivity has fallen down significantly because of small riceland. Their rice storage buckets, full of rice for the whole year in the previous year, became nearly empty the year after. The author got many responses like these:

> We used to never worry about rice for feeding our mouths. But it was really a nightmare for us when the shrimp farming established at this local area. Rice became the scarcest thing we could access.[17]
>
> We have never faced such a terrible situation as the current one. Last year, our rice storage in the house was full of rice. This year, it is nearly empty the whole year. Although it is not a profit activity, rice is indispensable to feed our mouths daily.[18]

In short, the loss and/or decrease of the access to arable land that resulted from the establishment of the shrimp farming project in the early 2000s put local households into positions of vulnerability because their traditional livelihoods were either alienated or insecure. However, it is also evident that the experiences of vulnerabilities among sampled households are different due to their differentiations in both human and natural resource access and livelihood development. Female heads of households, especially the poor that relied too much on these livelihoods for bringing up their whole family, found themselves under most adverse conditions; whereas women from MHHs at least could turn to the income from their male counterparts.

On the other hand, for households without access to forestland their inconvenience in terms of alienated traditional livelihoods and resultantly narrowed options for buffer livelihoods

---

[15] In-depth interview with Mr. and Mrs. Khan, a medium poor household, Aug. 11, 2011.
[16] Mrs. Thinh and Mrs. Hoa in a group discussion with female farmers, Jun. 21, 2011.
[17] Mr. Toan in group discussion with male farmers, Jun. 22, 2011.
[18] Mr. Nhat, a key informant, Jun. 19, 2011.

remained. In reality, because of the necessity to buffer the livings many local households, particularly female heads, people found surreptitious ways to collect fuel wood for home use or graze their cattle even though they perceived that such action is very dangerous and they could be fined by the forest manager, as in the case of Mrs. Phinh below.

> Now, natural forest is not allowed to access as before. But I need firewood for home consumption and sometimes sneakily access forest for firewood logging. Many women in this commune do this like me. Grazing my buffalo is another difficulty. I just got a fine by local committee because my buffalo entered an allocated forest for grazing grass.[19]

In addition, the changes in forest use narrowed down the area of pasture, which traditionally served as a place for cattle grazing and shelter. This in turn constrained women without forestland to remain and/or not to expand their animal production because of no shelter and grazing places, whereas to women with forestland their ruminants can still graze on their land.

In short, the changes in forest resource use brought out unequal and uneven consequences for particular households in practice. Being allocated forestland helped some households to have and maintain a buffer livelihood during times of agricultural stress and to improve their income. However, for other households, especially those with female heads, the associated absence of male labor in their households and their weak financial capacity prevented them from being allocated forestland.

15.4 CONCLUSION

This study presented the influencing role of institutional changes at a higher level on land distribution and use in practice.

These findings suggest to us that there exists a gap between policy and practice in land resource distribution and use. Local households, even if they live within a community, are not the same, with differentiated socioeconomic characteristics that differentiate their ability to access land for their own productive purposes. FHHs with their gender-specific disadvantages such as the lack of a male active labor are excluded from the forestland allocation process. As a result, local households, especially MHHs and FHHs, experienced differentiated vulnerabilities to climate-related agricultural water scarcity in the first place.

The findings also reveal that existing land-related policies are further putting many female heads in a position of marginalization that is exacerbated by climate-related changes. Therefore, development policies and national programs related to natural resources and particularly land have to shift their attention to incorporate social and gender differentiation in the formulation and implementation process. Otherwise, it is likely to exacerbate vulnerabilities of those affected, particularly female heads, rather than help them.

[19] Mrs. Phinh, poor female head of household in an in-depth interview, Aug. 6, 2011.

References

Agrawal, A., 2007. Forests, governance, and sustainability: common property theory and its contributions. Int. J. Commons 1 (1), 111–136.

Cooke, M., 2004. Na Kooliga' Gender Relations in Access To and Control Over Natural Resources in Rural Ghana. http://www.suhuyini.org/wp-content/uploads/Na-Kooliga-scriptie-Melissa-Cooke.pdf (accessed 19.01.14.).

Cornwall, A., 2007. Revisiting the 'gender agenda'. IDS Bull. 38 (2), 69–78.

Dang, K.S., Nguyen, N.Q., Pham, Q.D., Truong, T.T.T., Beresford, M., 2006. Policy Reform and the Transformation of Vietnamese Agriculture. Rapid Growth of Selected Asian Economies. Lessons and Implications for Agriculture and Food Security. FAO, Republic of Korea, Thailand and Vietnam, ISBN: 92-5-105509-2.

De Soto, Hernando, 2000. The Mystery of Capital: Why Capitalism Triumphs in the West and Fails Everywhere Else. Basic Books, New York, NY.

Do, Q.T., Iyer, L., 2008. Land titling and rural transition in Vietnam. Econ. Dev. Cult. Chang. 56 (3), 531–579.

FAO, 2010. National Gender Profile of Agricultural Households, 2010. Reported based on the 2006 rural, agriculture, and fishery census. Financed by the Swedish International Development Cooperation Agency (Sida), Sweden.

Hoang, T.S., 2009. Gains and Losses: Devolution of Forestry Land and Natural Forest. Retrieved on 14th February 2012 from Acta Universitatis agriculturae Sueciae. http://pub.epsilon.slu.se/2126/. 1652-6880, ISBN: 978-91-576-7419-7.

IPONRE (Institute of Strategy and Policy on Natural Resources and Environment, 2009. Ha Tinh assessment report on climate change. Hanoi, Vietnam. http://geodata.rrcap.unep.org/climate_change_report/HaTinh-Eng.pdf (accessed 15.11.11.).

Kirk, M., Nguyen, D.A.T., 2009. Land Tenure Policy Reforms. De collectivization and the Doi Moi system in Vietnam. IFPRI Discussion Paper 00927. This paper has been prepared for the project on Millions Fed: Proven Successes in Agricultural Development (www.ifpri.org/millionsfed).

Meinzen-Dick, R., Brown, L.R.F., Hilary, S., Quisumbing, A.R., 1997. Gender, Property Rights and Natural Resources. FCND Discussion Paper No. 29. International Food Policy Research Institute.

Menon, N., Rodgers, Y., Kennedy, A., 2013. Land Rights and Economic Security for Women in Vietnam. World Bank Working Paper. Washington, DC: World Bank.

MoNRE (Ministry of Natural Resources and Environment), 2011. Scenarios for Climate Change, Sea Level Rise in Vietnam. Hanoi, Vietnam. Vietnamese version. http://www.scribd.com/doc/84393795/K%E1%BB%8B-chb%E1%BA%A3n-B%C4%90KH-va-n%C6%B0%E1%BB%9Bc-bi%E1%BB%83n-dang-cho-Vi%E1%BB%87t-Nam-2011. (accessed 06.06.12.).

Naidu, S.C., 2011. Access to benefits from forest commons in the Western Himalayas. Ecol. Econ. 71, 202–210.

Ravallion, M., de van Walle, D., 2003. Land Allocation in Vietnam's Agrarian Transition. World Bank Policy Research Working Paper 2951.

Ribot, J.C., Peluso, N.L., 2003. A theory of access. Rural Sociol. 68 (2), 153–181.

Schlegar, E., Ostrom, E., 1992. Property-rights regimes and natural resources: a conceptual analysis. Land Econ. 68 (3), 249–262.

Sikor, T., Lund, C., 2009. Access and property: a question of power and authority. Dev. Chang. 40 (1), 1–22.

Sikor, T., Nguyen, Q.T., 2007. Why may forest devolution not benefit the rural poor? Forest entitlements in Vietnam's central highlands. World Dev. 35 (11), 2010–2025.

UNDP (United Nations Development Programme), 2013. The women's access to land in contemporary Vietnam. http://www.vn.undp.org/content/vietnam/en/home/library/democratic_governance/women_access_to_land_in_viet_nam.html.

Von Benda-Beckmann, F., von Benda-Beckmann, K., Spiertz, H.L.J., 1996. Water rights and policy. In: Spiertz, J., Wiber, M. (Eds.), The Role of Law in Natural Resource Management. VUGA, The Hague.

CONCLUDING SECTION

En Route to Effective Management of Natural Resources for Conservation and Livelihood Advances in Central Vietnam

S. Sharma*, G. Shivakoti[†,‡]

*WWF Nepal, Kathmandu, Nepal †The University of Tokyo, Tokyo, Japan ‡Asian Institute of Technology, Bangkok, Thailand

16.1 INTRODUCTION

Vietnam, positioned on the Indochinese peninsula, is among the wealthiest countries in terms of rich biodiversity and vast natural resources. The country is basically "S" shaped and has the Red River Delta and highlands in the north, and Central Mountains and coastal lowlands, which have forested hills and tropical lowlands terrain. Central Vietnam has rugged mountains, densely covered forest, and fertile soil along with timber, hydropower, and rich minerals. There are diverse people settled in these areas whose livelihood is dependent on natural resources, mainly forest, land resources, and water; up until recently this dependency was endangered by a number of disturbances.

The purpose of this chapter is to comprehend and analyze the diverse concerns in natural resources discussed throughout the earlier 15 chapters in this book, with implications to livelihood, economic dynamics, and unfolding the dynamism of efforts forwarded to reduce these threats with future implications for sustainability. This chapter is divided into four sections: land-use dynamisms after economic reforms, reconciling traditional knowledge and science in resource management, policy and institutions in resource management, and future implication of Vietnam's resource management into emission reduction programs in the first, second, third, and fourth sections, respectively.

16.2 MAJOR ISSUES IN CENTRAL VIETNAM

Communities in Central Vietnam are dependent on natural resources. Mostly, those residing at uplands and midlands are dependent on agriculture and forestry while communities at coastal areas are reliant on fisheries for their livelihoods. Especially, coastal areas have the interference of water resources both from freshwater rivers and the sea to form a special brackish water ecosystem with high biodiversity, which forms an integral system of livelihood security for fishing communities. Ever since rapid development activities were undertaken, coastal urbanization is increasing. This process of hasty urbanization has affected both the ecosystem and socioeconomic development. This situation is further aggravated by global climate change impacts. The frequencies of natural hazards are increasing concomitantly with change in storm intensity, erratic precipitation, temperature fluctuations, and floods, which have seriously damaged agriculture and fisheries. There have been incidences of soil and land degradations, accelerated by climate change.

The fishing communities and subsistence farmers are the most affected ones, experiencing silt depositions in the farmlands and fish mortality due to polluted water and siltation. Some fishing communities also believe these events, explicitly floods, bring in new fish species while others believe these events damage their fishing gear. Both ways, natural hazards have a strong influence on their livelihoods. However, the local communities are aware of climate change impacts on resources and ultimately its influence on their livelihood.

Likewise, although Central Vietnam is the most forested region of Vietnam and rich in biodiversity and endemism, it is threatened by biodiversity loss, deforestation, and land-use changes. Lowland forests of Central Vietnam have been replaced by fast-growing exotic species. "The natural tropical forests are fragmented while the remaining forest patches are isolated with plantation forests representing a barrier for the spread of most wild plants and animal species." Central Vietnam is experiencing a situation of biodiversity loss and degradation at rapid rate, while the documentation of these species is quite minimal.

16.3 SOCIOECONOMIC DYNAMISM AND LAND-USE CHANGES IN CENTRAL VIETNAM

Vietnam's reunification in 1975 was followed by relocating shifting cultivators into forest edges in the central midlands and encouraged into permanent settlements and stable agriculture. Synchronously, provincial administration reassigned natural resource management rights to state forest enterprises (SFEs) thereby eliminating private rights over forest. This led to mixed cultivation practices among local communities; some continued shifting agricultural practices while very few shifted into permanent agriculture. In late 1986, Vietnam launched an economic renovation policy, *doi moi*, intended to devise a fair market economy. There were major changes, explicitly in agriculture, achieved through "market liberalization and integration into the international economy." In Central Vietnam, there was easy accessibility to high-yield crops, technology, loan facilities, and efficient service delivery from a government that encouraged local cultivators toward agriculture intensification.

This was followed by decentralization policies launched to transfer management rights from a central authority to provincial and district bodies, especially for forestry development degraded forestland allocation (FLA); a policy that turned out to be the strongest foundation in forest decentralization. In Central Vietnam, this policy focused on restoring plantations over barren land and degraded forest. Many interventions by SFEs also focused on restoring degraded forest were launched laterally through acacia and rubber plantations to increase household benefits. Simultaneously, short-term contracts were signed with communities for undertaking patrolling and regular monitoring of forests. By 2003, regional administration in Central Vietnam allocated natural forest under SFE to locals, and this step strengthened the role of local communities as implementing partners for forest allocation mechanisms. Nevertheless, forest cover dynamism was high after 2001, though some scholars conclude that the forest cover increment occurred from 2001 to 2004, while others see this as process errors. During that period, the conversion rate of damaged bare land and degraded forest to oil and palm tree estate plantation was significantly higher than earlier periods. A hasty farmer's decision toward plantation crops was supported by *doi moi*. Principally, policy and socioeconomic settings in northern Vietnam were responsible for this change. State sponsored timber harvesting and the consequent household logging in the forest on the verge of being allocated was the primary reason for degradation in 2001. On the contrary, loan facility, strong extension from the central administration, and the success demonstrated in northern Vietnam were influencing factors for household land-use change. Case studies from this book also have identified that gearing up markets, slow implementation of forest allocation, and weak enforcement of forest protection encouraged some local households to access forests more and resume shifting cultivation in the forest, further creating vigor at local landscapes. High economic value of acacia and regional market extensions for pulp and rubber also triggered land transitions beyond forest boundaries. This could be linked with the ease of forest rights use within plantation agreements with SFEs. These factors may affect the sustainability of forest conservation in the near future in Central Vietnam.

Size of the available land also triggered land transition to plantations; explicitly, households with higher landholding proportions tend to convert more for higher income from potential commercial plantations. Higher landholding households are at little risk of failure from new land-use transitions, potentially motivating them to expand land-use practices to create new income sources. On the contrary, smaller landholding households are at risk of market failures because they cannot afford expanding available land for additional income. This has discouraged small landholders to undertake new technologies and available resources; quite the reverse, they are inclined more toward subsistence agriculture. Individuals with rich landholding set aside portions of land for soil enrichment and restoration. Market scopes for high-value crops tend to steer these landholders to use this reserved land, which provides them with insurance if the market fails. Simultaneously, high landholders who are likely to have higher welfares than smaller landholders can help maintain inputs for soil fertility and soil gradient, respectively, through crop species rotations. On the other hand, households settled for a longer period of time are endowed with cumulative experience on higher agricultural productivity, improved knowledge on resource conditions, domestication of crops, and market situation, and they embrace new land uses swiftly. Cases in the book reveal that most of these land-use changes are linked with policy initiatives, explicitly the National Forest Assessment (NFA) that was focused on allocating degraded natural forest to individual households and communities. Internal forest dynamism augmentation coincided with NFA polices in Central Vietnam.

VI. CONCLUDING SECTION

The cases in the book disclose that deforestation and degradation were the result of land speculation, mostly by *Katu* and *Kinh* indigenous communities. Weak forest conservation policies together with high forest dependency of these communities triggered forest allocation. In the meantime, there were land-use changes outside forest due to user's land-use decisions as intermediated by a high-end level of awareness on the market possibility of plantation trees and cash crops, loan facility, and strong extension services. These were further underpinned by technology and policy influences, exclusively markets, road development, and demography.

16.4 SOCIAL IMPACTS OF FOREST LAND ALLOCATION POLICY

Since the inception of FLA in Central Vietnam, the program has successfully intensified and restored forestland together with enhancing local communities' awareness toward forest consumption in a balanced way. But queries persist on whether FLA has influenced the livelihood of forest-dependent communities or, conversely, created potential risk toward livelihood transformations.

The book's chapters disclose that resource-poor households were less benefited through FLA in terms of quantity and quality in comparison with resource-rich households with stronger societal positions. This indicates an imbalanced race in terms of FLA among diverse social status. As a result, FLA gives the impression of having not helped resource-poor individuals to overcome their existing poverty status while "further widening the gap between rich and the poor." According to a case study from Thuong Nhat Commune in Central Vietnam, the forestland was open access for individuals with an interest to undertake and invest in reforestation program, but resource-poor households were least interested until they realized the value of forest. And by that time fertile and productive forests were already allocated to powerful state stakeholders. This gives the impression that FLA was applicable to households with the capability to invest with good social networking correspondingly. Similarly, rich households had more than one piece of land, while the resource-poor had only one; during catastrophic situations poor households would sell their plots to the rich and become landless quite easily. This has again widened the gap between the rich and the poor. Likewise, FLA allows land-use rights for 50 and 20 years for forestland and agricultural land, respectively, whereby no outsider could cultivate or harm the particular land. As such, local authorities cannot reallocate the land once it has been provided to a particular household. With the population growing at an alarming rate it is obvious that a young couple have no land left to cultivate. With no options left, young couples are compelled to reclaim land that was once owned by their forefathers. Though considered illegal, individuals continue to have the expectation that the state may allocate the land further. Even though the government has allocated forestland to individual households, encroachments in forest areas are uninterrupted and a conflict situation may soon break out between the state and local households.

FLA has encouraged native communities to undergo sedentary cultivation. However, this has forced communities to cultivate the same piece of land now and then; lowering the regeneration capacity of the soil. This has encouraged farmers to abandon native species that do not perform well in these less fertile soils and tend to opt for improved varieties with higher yields and a shorter harvest cycle. This has also simultaneously increased farmers' motivation toward plantation crops, as native crops responsible for food security are yielding less and plantations are offering more. This trend could envision a situation of local agrobiodiversity loss and reduced self-reliance over food in near future. This will possibly render a contradictory rationale between ecological conservation and economic development in Central Vietnam. Soon after FLA implementation, changes in household income, better living standards, household access to resources, increased networking among stakeholders, and increased women's voice were significant changes observed. Changes are also observed in terms of knowledge and skills development and overall livelihood sustainability. However, case studies also reveal increased "landlessness and unemployment," as mentioned above. This situation is likely to flourish concomitantly with population growth.

However, in terms of economic benefit from perennial plantations the high returns have braced farmers to escape poverty. But, as larger numbers of farmers are involved in perennial plantations followed by provincial authority focusing on expanding plantation areas, there is always a high possibility of market price reduction. This is likely to have severe effects as no attention toward future demand and supply of these plantation products is considered. These situations may "replicate the situation of oversupply" as Vietnam once suffered from excess sugarcane and cashew nuts (Tam, 2008). As perennial plantations are mostly synonymous with monoculture, there is a strong possibility of notorious pests infesting entire plantation populations. This would put farmers in a more vulnerable situation.

Land is a fundamental asset of natural resource-dependent communities in developing countries and, hence, a secure land tenure arrangement can enhance socioeconomic well being of these communities. Land is an important source for food and income, and plays a major role in sustainable livelihood. This is especially vital for women, as women's contributions to household income are widely argued. Realizing the importance of rights and tenure arrangement, the government of Vietnam has a number of policy documents to enable the environment for a successful economy and a good livelihood. As such, the rural people's access to land has significantly increased. On the other hand, access to land is not necessarily an easy mechanism, whereby land access is discriminated in terms of social and gender status. Case studies from the book have shown male-headed households to have obtained a larger share of cropland versus female-headed households. Arguably, household size of male-headed households is believed to be larger than that of female-headed households, and as Vietnamese culture considers males receive family property inheritance, males possibly get larger landholdings during the forest allocation process. In short, there exists a huge gap between FLA policies and reality laid in practice during land allocation and use. Many local households are discriminated against in terms of socioeconomic characteristics that affect their ability to access land for their own productive purposes. Poor and female-headed households have specific disadvantages against resource-rich and male-headed households. They have experienced a diverse level of vulnerabilities and scarcities.

16.5 RECONCILING SCIENCE AND TRADITIONAL KNOWLEDGE IN RESOURCE CONSERVATION

Although Vietnam has decentralized forestland to communities for 50 years for conservation and forest enrichment, most of the indigenous tree species endemic to Vietnam are currently lost while others are listed in the "Red Book" as threatened species. Some of these species have higher economic and esthetic value but are not taken into consideration due to a lack of knowledge and technology accessibility. The book chapter on propagation of *Scaphium lychnophorum* and *Baccaurea sylvestris* motivates the integration of scientific research together with legal rights to improve the livelihood of local people through cultivating valuable trees that are on the verge of extinction. With the particular example of these two species that are unable to germinate naturally, they could be rejuvenated through hot water treatment and simple grafting.

Ecologists are focused on identifying indicator species for environmental monitoring for biodiversity loss and forest condition changes. But traditionally local communities are aware of species that define forest conditions, and these species are often deliberated in a sensitive way to consider them as early warning signals toward ecosystem degradations. A major problem in developing indicator species in Central Vietnam is that usually these indicators are developed by researchers with less integration toward indigenous expertise. As such, local observational experience and historical context are not included in the indicator set. Hence, it has become difficult to obtain good indicators to define forest conditions and foresee better collaborative conservation and long-term forest management. However, communities' response on particular indicator species is highly reliant on the particular use of trees and access to them. Community-oriented indicators convey early warning for forest degradation and simultaneously forecast a cause-and-effect correlation flanked by indicator species and their incidence in an exact forest disturbance. At the same time seedlings may act as indicators for particular forest management practices, as it forecasts the progression of forest vigor if positive environments are available for seedlings growth. Likewise, the occurrence of seedlings of a particular species in heavily degraded forest may forecast the direction of forest type in the future. Some scholars recommended that forest monitoring initiatives using scientific methods could delay or hamper forest protection in developing nations if they had to deter resources away from basic forest management priorities. The confusions may be set by monitoring through forest indicator species. In this monitoring, local experience and knowledge of forest species is combined with quantitative statistics. This can monitor the forest while also increasing local awareness for conservation.

Agriculture land fragmentation and inappropriate cultivation practices are lowering food productivity and simultaneously increasing poverty in the highlands of Central Vietnam. Most of the productive forestland are converted to agriculture and residential areas; rendered a negative influence on vegetative cover pertaining to soil losses. In the process, the fertile topsoil is washed away, is deposited into water bodies and lakes, and ultimately creates negative impact on the environment and the economy. In this regard, soil erosion issues are the highest prioritized issue for policy makers. However, not much attention is given to watershed soil erosion models to target affected zones. Various soil erosion models may be used to evaluate total sediment deposited within a catchment to a river basin. The findings from the book state that the largest amount of soil erosion took place in 2007 and 2010, correspondingly, whereby $62.50\,t\,ha^{-1}$ was lost in totality. These losses were high in dry agriculture with slopes above 25 degrees. Applying these tools is reliable and supports the resource manager to make efficient decisions on building plans for land cover changes and identifying potential erosion.

VI. CONCLUDING SECTION

16.6 POLICY AND INSTITUTIONS

16.6.1 Community Forestry for Resource Management

Community forest management (CFM) is one of the more successful management models around the world. Local forest-dependent communities within this model have direct responsibility for forest conservation together with bundles of use and management rights. In Vietnam the forest management approach moved from centrally managed to community managed during the early 1990s. During the early 2000s, Vietnam had developed an institutional framework and related policies for CFM that concluded it was an authentic Vietnamese model for forest management. The anticipation of this model in Vietnam contributes directly to sustainable forest management while it improves the livelihood of forest-dependent communities. In Central Vietnam, this model was initiated as a pilot project in early 2000. "Thuy Yen Thuong Village, Loc Thuy Commune, Phu Loc District" were initially screened to undertake this model, whereby approximately 400 ha of forest area was delineated to these communities for management. Awake of 2009, "around 10,904.7 ha of forestland handed over to the communities for long term management for typically 50 years in Phu Loc, Nam Dong, A Luoi and Phong Dien districts" (TTHDARD, 2009).

Especially in Central Vietnam the type of forests allocated are protective and production forests allocated on the basis of household, household groups, and village groups; mostly concentrated in severely to moderately degraded forests. More than one model has been tested and applied in the entire country to capture diverse settings and construct benefit-sharing mechanisms. In the particular CFM model in Central Vietnam, rights to access, use, management, and exclusion were given with timber harvesting that exceptionally required certain conditions to harvest. The communities involved are ethnic minorities residing in remote areas and using these rights prior to the actual application of the CFM concept. Local ethnic practice allowed these communities to access any land, as they believed land to have been bestowed by nature, and that conflicted with CFM policies of demarcating forest boundaries. As a result, there is far less participation from these communities in the process of forest allocation and also people do not take part in any meetings arranged by the local government's forest development board. They did not participate in regular forest patrolling activities either, while the frequency of nonethnic households participating in patrolling also reduced simultaneously with the passing of time as most of them demanded salaries. In addition, though polices allowed locals to benefit from forest products equitably and encouraged participation, these were least achieved in the fields.

More than one model has been tested and applied in the entire country to capture diverse socioecological settings, participation, and solve issues accordingly. But this has created confusion during planning and implementation in a very real sense. However, models with clearly mentioned legal and tenure arrangements, focused on internal governance dynamics, acknowledging local knowledge of forest conservation and cost incurred in the process, and the relationship with service agencies have outdistanced other models. A study mentioned in the book reveals that for CFM models to be sustainable, benefit-sharing polices should be based on profitability from each forest resource and the efforts put into it. In the process, Central Vietnam is required to create legal corridors for communities to get access to information on legal policy provisions on their tenure rights; the local government agency has the primary role of disseminating and transferring legal documents to the community to promote informed decision making. This should be followed with establishing and strengthening an internal governance system to increase participation in planning and monitoring. Apart from extension services, regular patrolling of the forest can help to understand the resource better and to construct livelihood models developed through this knowledge.

16.6.2 Comanagement: An Approach to Conflict Resolution

Prior to the 1990s, forests in Central Vietnam were managed under the state government, which was followed by a decentralization process after the inception of *doi moi* through multistakeholder involvement in forest conservation and use. In the process, production and special-use forest were devolved to local communities while protected forest remained under the custody of state government. These allocated forests were poorly degraded and rendered subtle benefits to local communities. Thus, local people continued accessing restricted timber and nontimber forest products (NTFPs) from protection areas, too. Soon, a dispute occurred between the communities' and officials from protected areas granted to conserve the area through prohibiting local people in the core area. This preference difference ignited conflicts leading to further degradation of protected areas with rampant incidences of forest fires, swiddening, haphazard hunting, and illegal logging, and ultimately led to epidemics of species extinction. Identifying the restriction of access and use rights in protected areas as a primary cause of problems, a mechanism of comanagement was implemented in Central Vietnam. The basic ideology was to involve local people in the planning process and open up opportunities for the community's access in a buffer zone while leaving behind a core area for absolute conservation. Simultaneously, a capacity building package to sustainably manage forest comprehended diverse logical steps of carrying out participatory resource inventory, long-term planning, rules formation, and resource extraction to set up steps for sustainable forest management through "learning by doing." Property rights and secure tenure arrangement were the key to achieve success in comanagement in Central Vietnam, which provided incentives to local communities for technical and financial investment.

16.7 SOCIAL TENSIONS OF CENTRAL VIETNAM'S RESETTLEMENT PROGRAMS

Resettlement is the intricate course that focuses on "compensation, resettlement, livelihood restore, and natural resource access change." This is an interesting topic as Central Vietnam is home to several hydropower plants and it has dislocated communities residing in the vicinity. Hydropower is synonymous with dam construction and intricacy arises as most affected people are ethnic minorities. Regardless of how the administration put together the specific resettlement relocation strategy and displacement package to lessen adverse effects on the communities, a lot of complications exist that affected communes have to face.

The Vietnamese resettlement legal provision includes "compensation, displacement and resettlement, and additional livelihood" upkeep next to resettlement at the new site. The respective resettlement scheme is constituent with a number of rules formulated on the basis of general state rules and investor's financial and technical capacity. Nevertheless, most of these schemes are concentrated on infrastructure development and construction of settlements for displaced households. Mostly in Central Vietnam, according to cases described in the book, this process is slow and tedious, often leading to discontent. In terms of access to natural resources, the quantity and quality of forest products are observed as significantly reduced subsequent to resettlement. These conditions have been argued to have further deteriorated and marginalized the livelihoods of affected communes. Furthermore, no model directed toward alternative livelihood exists in Central Vietnam. Most of the individuals had to transform their livelihood strategies and acclimatize themselves to a new environment while their knowledge, skills, and education are low. Food insecurity has arisen while people are unemployed and social evils have drastically increased.

Hydropower growth has profits for developing markets that have a high plea for hydroelectricity. The rise in the sum of hydroelectric schemes that hints at the escalation of households that are impacted negatively. Dislocation and relocation often changes the living situation and means of support of affected communes. Other problems cited are lack of or decrease in water access, and individuals do not have sufficient drinking water nor do they engage in agricultural cultivation. "The total land area per household and efficiency of land use fell sharply." On the other hand, households with a land certificate increased from 0 to 100 percent before and after resettlement, respectively.

16.8 POLYCENTRIC APPROACH TO CLIMATE CHANGE AND EMISSION REDUCTION

Natural resources management in Vietnam is built on the foundation of sustainable development and environmental protection. Nonetheless, Vietnam lost around half of its forest from 1943 to 1990 (Nguyen, 2013), which meant that Vietnam had the second highest deforestation rate in the world; further aggravated by climate change. The 13th Conference of the Parties (COP 13) identified Vietnam as one of the five countries to be affected by climate change. The consequences of climate change have been prominent through reduced forest and agriculture productivity and a decline in the quantity and quality of water resources; simultaneously intimidating economic gains mostly gained through the *doi moi* policy. The country has developed strategies to address multifaceted issues in climate change but these strategies often are not deliberated as demands of local communities. Vietnam is 1 of the 13 countries selected for future emission reduction programs. In the process, Vietnam is preparing itself through devising institutional arrangements and organizational reforms. But as Ostrom (2009) argues, singularities of legal provisions are not sufficient to render trust among local stakeholders and confirm collective action in a transparent and comprehensive manner to address climate change and global warming issues. In this section, we try to summarize multifaceted attributes likely to affect future emission reduction programs.

16.8.1 Informed Decision Making

Reducing Emissions from Deforestation and Forest Degradation (REDD+) calls for the effective participation of diverse stakeholders for transparent resource governance information on ways to mitigate climate change and adaptation interventions, and economic and social safeguards are being incorporated in the policy. A number of policies strengthen participation and an informed decision-making process. The Constitution of Vietnam, Vietnam's Law on Environmental Protection, Law on Land, Vietnam's Agenda 21, the Forest Protection and Development Law, and others specify organizations and individuals to be involved in environmental protection management through the right to information. As indigenous communes reside adjacent to natural resources, their existence and rights have been recognized legally. However, there are overlaps among polices that often create confusion through inconsistent definitions and interpretations.

16.8.2 Planning Process

Vietnam's climate change strategies are focused on retrieving impacts of climate change and then developing plans for obtaining a low carbon economic trajectory. The Ministry of Natural Resources and Environment (MONRE) is responsible for overall climate change planning while the Ministry of Agriculture and Rural Development (MARD) simultaneously develops action plans for adaptations in agriculture while other sectorial departments make plans accordingly. Locally, committees at each level commence land management in their dominions, prepare effective plans, and submit it to an upper hierarchy level for approval. These are then combined to develop a national plan. In a way, Vietnam's governance system determines foundational mandates of actors from the local to the national level. However, there exists turf between departments and diverse authorities creating confusion and rules overlaying are often considered difficult to understand and implement.

16.8.3 Property Rights and Benefit-Sharing Mechanisms

"The state owns all forests in Vietnam, which are then allocated to households and organizations for short term or long term benefits." In the context of selling and transferring greenhouse gases, it is considered as one of the forest products that could be marketed as any form of forest by-product; but carbon marketing exclusively requires prime ministerial approval and this may defy the REDD+ scheme. The fact that current rights include carbon rights or not may pose an additional challenge to REDD+; according to Covington et al., "uncertainty surrounding land title is significant preconditions for REDD+ scheme." To address this situation, legal certainty on land titles is essential. This may have consequences to ethnic minorities where no legal title is given and a huge gap is seen between de jure and de facto rights. Particularly, legal and institutional challenges will only be ensured if the rights are managed equitably; otherwise, competing claims from the ethnic minorities are likely to arise and many communities may lose access to forest resources.

Equitable benefit distribution systems may be understood to initiate from the Cancun Agreements as a precursor to guidance and several safeguards. Particularly, equitable benefit distribution aggravates benefits to directly involved communities from REDD+, contributing to overall forest management, poverty eradication, and sustainable development. It also delivers effective interventions and permanence by creating incentives for carbon sequestration through results-based payment. Vietnam does have experience in such results-based equitable benefit distribution systems through Payment for Environmental Services (PES) programs. Lessons from PES can aid in research and identifying issues on REDD+ implementation that significantly includes degrees of compensation required to fully engage community into PES schemes along with the management and timely disbursement of funds; as such it may serve for a future REDD+ sharing mechanism.

16.9 RECOMMENDATIONS

Built on the conclusions from the book, we forward the following recommendations.

(1) Regardless of the fact that research plays a major role in generating ideologies and addressing problems, it is often not considered in the policy formation process. Local beneficiaries from Central Vietnam can play an active role in identifying problems and sorting ways to solve them. They should be brought into the planning process.

(2) In the context of ongoing land-use changes subsequent to FLA, attempts to promote small-scale industries could likely generate off-farm job opportunities, which in turn would reduce pressure on the forest. This should be concluded with widening the commune's role in decision making and eventually developing the provincial government's capacity toward forest allocation responsiveness.

(3) Owing to the fact that agriculture land transformed to estate plantation and severely damaged income through market fluctuation prior to implementing FLA, an attempt should made to select appropriate species for plantations, providing tenure assurance and establishment of plantation markets that are profitable for resource-poor farmers.

(4) Rather than climate change research focusing on the impact of global warming, attempts to forward alternative livelihood solutions for affected communities and document adaptation practices should be undertaken locally. Additional work on improved climate knowledge network formation could render early warning systems that effectively help communities to prepare themselves for the adverse effect.

(5) Apply scientific tools that are reliable and can be useful for resource managers and policy makers in setting up scenarios. Hence, it is recommended to refer to scientific studies prior to undertaking policy formulations and management approaches. Regional and knowledge networks need to be consulted for scientific solutions, to build science-based approaches, and to reduce misperceptions over problems occurring at an ecosystem scale.

(6) Displacement and resettlement should have a good compensation policy linked with resource access and rights to water and other resources. It should be made sure that new resettled sites must have natural resources similar to the earlier sites. Compensation should not only focus on building houses but also on access to resources. Additionally, awareness of rights should be raised among all the affected communities to ensure compensation has been calculated adequately. The infrastructure development owner should be required to set up a fund to support affected communities.

(7) To stabilize and enhance the quality life of affected households, the resettlement program needs to ensure natural resources access by affected people, especially the quantity and quality of natural resource have to be equal to or better than before resettlement. Livelihood support should be continued for a long time and the

alternative livelihood models need to bring a commune to the resettlement area before people have to move on to a new place based on their knowledge and customs.

(8) Land-use planning significantly contributes to socioeconomic development and environmental protections. To anticipate efficient land-use planning, it is crucial to analyze diverse development scenarios with different settings, explicitly including the physical environment and socioeconomic criteria. Particularly, environmental factors that concomitantly affect development and livelihood of people need to be addressed.

(9) The role of local stakeholders in forest governance needs to be enhanced. This should be followed by developing an information-sharing mechanism that clearly defines stakeholder's rights and duties and incentives toward forest management. Local people's participation can be increased through awareness and economic incentives.

(10) Vietnam is legally ready for REDD+ interventions, but this necessitates additional considerations in legal supervisory measures and regulatory improvements for cooperative solutions, typically from the ground up.

References

Nguyen, H.H., 2013. Transition to sustainable forest management and rehabilitation in Vietnam. In: Paper Presented at International Symposium on Transition to Sustainable Forest Management and Rehabilitation: The Enabling Environment and Roadmap. 23–25th October, Beijing, China.

Ostrom E., 2009. A Polycentric Approach for Coping With Climate Change. World Bank Policy Research. Working Paper No. 5095.

Tam, L.V., 2008. The Forest Land Use and Local Livelihood. Case Study in LocHoa Commune, PhuLoc District.

TTHDARD (Thua Thien Hue Department of Agriculture and Rural development), 2009. Abstract Report on Forest Allocation Proposal of Thua Thien Hue Province in Period 2010–2014.

Appendix A

No.	Tree Species	Plant Family	Counts			Basal Area (m² ha⁻¹)	Biomass (kg/tree)	Importance Value Index
			Mature	Sapling	Seedling			
1	*Acronychia pendunculata* (L.) Miq	Rubiaceae	8	2	4	0.0151	65.61	1.40
2	*Actinodaphne pilosa* (Lour.) Merr.	Lauraceae	1			0.0039	14.75	0.19
3	*Actinodaphne* sp.	Lauraceae	1			0.0084	23.81	0.21
4	*Aglaia* sp.	Meliaceae		2				
5	*Aidia oxyodonta* (Drake) T.Yamaz.	Rubiaceae	5	6		0.0143	78.61	0.98
6	*Alangium kurzii* Craib.	Alangiaceae	1	2		0.0032	4.72	0.19
7	*Alangium ridleyi* King.	Alangiaceae	4			0.0295	139.59	0.78
8	*Amesiodendron chinense* (Merr.) Hu	Sapindaceae	18	1		0.0338	184.00	3.56
9	*Antidesma hainanense* Merr.	Euphorbiaceae	21	17	38	0.0120	42.20	3.39
10	*Archidendron clypearia* (Jack.) Nielson.	Mimosaceae	1	2	6	0.0042	9.19	0.20
11	*Archidendron* sp.	Mimosaceae	1			0.0207	84.77	0.23
12	*Artocarpus rigidus* A. Rich.	Moraceae	74	5	148	0.0279	149.83	13.37
13	*Artocarpus* sp.	Moraceae	2			0.0085	36.16	0.41
14	*Artocarpus styracifolius* Pierre.	Moraceae	4	1		0.0602	431.59	1.31
15	*Barringtonia macrostachya* (Jack.) Kurz.	Lecythidaceae	54	12	11	0.0385	201.36	11.17
16	*Barringtonia* sp.	Lecythidaceae	4	1		0.0550	326.54	1.14
17	*Breynia fruticosa* (Linn.) Hook. f.	Euphorbiaceae	2	1		0.0224	104.13	0.48

Continued

No.	Tree Species	Plant Family	Counts			Basal Area (m² ha⁻¹)	Biomass (kg/tree)	Importance Value Index
			Mature	Sapling	Seedling			
18	*Bridelia balansae* Tutch.	Euphorbiaceae	1	1		0.0347	229.88	0.27
19	*Calophyllum dryobalanoides* Pierre.	Clusiaceae	2			0.0050	14.23	0.27
20	*Calophyllum* sp.	Clusiaceae	2			0.0282	258.30	0.50
21	*Canarium album* (Lour.) DC.	Burseraceae	47	10	13	0.0250	127.45	8.21
22	*Canarium bengalense* Roxb.	Burseraceae	8	4	1	0.0960	593.18	3.07
23	*Canarium* sp.	Burseraceae	2			0.0042	15.04	0.27
24	*Canarium tramdenum* Dai. *ex* Yakovl.	Burseraceae	1			0.0058	24.62	0.20
25	*Canthium dicoccum* (Gaertn.) Merr.	Rubiaceae	1			0.0084	35.31	0.21
26	*Castanopsis* sp.	Fagaceae	73	5	8	0.0312	175.22	12.60
27	*Cinnamomum cambodianum* Lecomte.	Lauraceae	1	1		0.0161	87.90	0.22
28	*Cinnamomum camphora* (L.) H. Karst.	Lauraceae	1	1		0.0084	23.81	0.21
29	*Cinnamomum cassia* Nees *ex* Blume.	Lauraceae	4	1	2	0.0588	366.84	1.18
30	*Cinnamomum hainanensis* Merr.	Lauraceae	4	3		0.0482	272.24	1.08
31	*Cinnamomum obtusifolium* (Roxb.) Nees.	Lauraceae	19	10	16	0.0103	43.54	3.38
32	*Cinnamomum* sp.	Lauraceae	2			0.0461	279.15	0.66
33	*Cratoxylon ligustrinum* Blume.	Hypericaceae		1				
34	*Cratoxylon pruniflorum* Kurtz	Hypericaceae	9	3	3	0.0220	113.27	1.77
35	*Croton roxburghii* Balak.	Euphorbiaceae	1			0.0287	103.68	0.25
36	*Croton* sp.	Euphorbiaceae	3			0.0462	259.45	0.72
37	*Croton tiglium* L.	Euphorbiaceae	6	9	46	0.0046	10.92	1.06

No.	Tree Species	Plant Family	Counts			Basal Area $(m^2 ha^{-1})$	Biomass (kg/tree)	Importance Value Index
			Mature	Sapling	Seedling			
38	*Cryptocarya lenticellata* Lecomte.	Lauraceae	7	1		0.0470	318.84	1.83
39	*Cryptocarya* sp.	Lauraceae	1	1		0.0215	144.62	0.24
40	*Demos* sp.	Annonaceae	5	3	5	0.0208	77.42	0.91
41	*Dillenia scabrella* Roxb.	Dilleniaceae	11	20	37	0.0067	21.88	1.85
42	*Diospyros apiculata* Hiern	Ebenaceae	2	1		0.0063	20.15	0.28
43	*Diospyros eriantha* Champ. *ex* Benth.	Ebenaceae	2	1	5	0.0074	32.04	0.41
44	*Elaeocarpus apiculatus* Masters	Elaeocarpaceae	4		1	0.0118	57.06	0.85
45	*Elaeocarpus dubius* DC.	Elaeocarpaceae	4	2	3	0.0134	74.08	0.75
46	*Elaeocarpus griffithii* (Wight) A Gray	Elaeocarpaceae	10	3	4	0.0153	63.79	1.73
47	*Elaeocarpus nitentifolius* Merr. & Chun	Elaeocarpaceae	1			0.0029	4.28	0.19
48	*Elaeocarpus* sp.	Elaeocarpaceae	4	1		0.0111	40.63	0.60
49	*Endospermum chinense* Benth.	Euphorbiaceae	34	2	7	0.0848	585.23	10.72
50	*Enicosanthellum plagioneurum* (Diels) Ban	Annonaceae	7			0.1771	1206.21	4.25
51	*Engelhardtia spicata* Lesch *ex* Blume.	Juglandaceae	3	1		0.0330	260.41	0.67
52	*Euodia lepta* (Spreng.) Merr.	Rutaceae	13	4	40	0.0099	39.49	2.23
53	*Eurya ciliate* Merr.	Theaceae	4			0.0059	24.03	0.68
54	*Eurya japonica* Thunb.	Theaceae	3	1		0.0152	70.34	0.54
55	*Eurya nitida* Korth.	Theaceae	2			0.0400	217.30	0.44
56	*Eurycoma longifolia* Jack.	Simaroubaceae	12	7	25	0.0059	17.71	2.15
57	*Ficus glandulifera* Wall.	Moraceae	1			0.0472	310.20	0.30
58	*Ficus hispida* (FH) Linn.	Moraceae	1			0.0067	9.73	0.20

Continued

No.	Tree Species	Plant Family	Counts			Basal Area (m²ha⁻¹)	Biomass (kg/tree)	Importance Value Index
			Mature	Sapling	Seedling			
59	*Ficus* sp.	Moraceae	26	2	5	0.0174	68.83	4.06
60	*Ficus trivia* Corner	Moraceae	1			0.0510	225.30	0.31
61	*Ficus vasculosa* Wall. *ex* Miq.	Moraceae	4	1		0.0074	24.65	0.57
62	*Garcinia cochinchinensis* (Lour.) Choisy	Clusiaceae	45	14	90	0.0173	88.40	7.38
63	*Garcinia cowa* Roxb.	Clusiaceae	1			0.1675	923.72	0.58
64	*Garcinia* sp.	Clusiaceae	6			0.0193	88.40	1.27
65	*Garuga pinnata* Roxb.	Burseraceae	76	13		0.0233	111.27	11.94
66	*Glochidion eriocarpum* Champ. *ex* Benth.	Euphorbiaceae	20	15	45	0.0131	80.15	3.35
67	*Glochidion zeylanicum* (Gaertn.) A. Juss.	Euphorbiaceae		1				
68	*Gonocaryum maclurei* Merr.	Icacinaceae	13	3	13	0.0144	43.38	2.00
69	*Gironniera* sp.	Ulmaceae	5	4		0.0055	20.49	0.75
70	*Gironniera subaequalis* Planch.	Ulmaceae	83	44	42	0.0196	88.42	12.78
71	*Harmandia mekongensis* Balll.	Olacaceae	3	1		0.0233	136.32	0.60
72	*Hopea pierrei* Hance	Dipterocarpaceae	1	1	1	0.0603	418.75	0.33
73	*Hopea* sp.	Dipterocarpaceae	2		1	0.0045	12.56	0.39
74	*Horsfieldia amygdalina* (Wall.) Warb.	Myristicaceae	18	2		0.0325	183.25	3.87
75	*Ilex megistocarpa* Merrill, E.D.	Illiciaceae		1				
76	*Illicium griffithii* Hook. f. & Thoms.	Illiciaceae	7	5	8	0.0093	40.50	1.21
77	*Ixonanthes reticulata* Jack	Ixonanthaceae	3			0.0069	26.34	0.36
78	*Knema conferta* (King) Warb.	Myristicaceae	46	11	10	0.0143	57.54	6.92
79	*Knema pierrei* Warb.	Myristicaceae	51	13	21	0.0167	74.73	8.18

No.	Tree Species	Plant Family	Counts			Basal Area (m² ha⁻¹)	Biomass (kg/tree)	Importance Value Index
			Mature	Sapling	Seedling	(m² ha⁻¹)	(kg/tree)	
80	*Litsea euosma* W.W. Sm.	Lauraceae	1			0.0050	17.76	0.20
81	*Litsea glutinosa* Lour. Rob.	Lauraceae	32	10	9	0.0119	46.25	4.88
82	*Litsea lacilimba* (Hance) A. Camus	Lauraceae	3	2		0.0150	67.81	0.66
83	*Litsea pierrei* Lecomte.	Lauraceae		2				
84	*Litsea sebifera* Pers.	Lauraceae	1	2		0.0042	8.20	0.20
85	*Litsea* sp.	Lauraceae	2			0.0526	235.25	0.50
86	*Litsea verticillata* Hance.	Lauraceae	2	1		0.0141	49.97	0.32
87	*Lithocarpus amygdalifolius* (Skan) Hayata	Fagaceae	71	5		0.0330	171.81	12.51
88	*Lithocarpus farinulentus* (Hance) A. Camus	Fagaceae	2			0.0065	24.44	0.28
89	*Lithocarpus* sp.	Fagaceae	2			0.0233	135.23	0.48
90	*Machilus* sp.	Lauraceae	2	1		0.0045	16.36	0.27
91	*Madhuca pasquieri* (Dubard) H. J. Lam	Sapotaceae	6			0.0315	175.23	1.32
92	*Mallotus apelta* (Lour.) Müll. Arg.	Euphorbiaceae	13	3		0.0212	106.53	1.97
93	*Mallotus paniculatus* (Lam.) Mull. Arg	Euphorbiaceae	1			0.0098	34.16	0.21
94	*Mangifera minitifolia* Rose.	Anacardiaceae		2				
95	*Melanorrhoea laccifera* Pierre.	Anacardiaceae	21	20	41	0.0175	81.45	4.03
96	*Microdesmis caseariaefolia* Planch. ex Hook.	Sapotaceae	7	1		0.0378	208.56	1.80
97	*Mimusops elengi* L.	Sapotaceae	7			0.0378	208.56	1.80
98	*Mischocarpus fuscescens* Blume.	Sapindaceae	2	1		0.0179	64.06	0.46
99	*Morinda citrifolia* (Noni)	Rubiaceae	49	15	4	0.0174	72.81	8.18

Continued

No.	Tree Species	Plant Family	Counts			Basal Area (m²ha⁻¹)	Biomass (kg/tree)	Importance Value Index
			Mature	Sapling	Seedling			
100	*Morus alba*	Moraceae	17		1	0.0233	113.21	3.36
101	*Nephelium chryseum* Blume.	Sapindaceae	2	1		0.0073	28.23	0.41
102	*Nephelium lappaceum* L.	Sapindaceae	1			0.0523	297.84	0.31
103	*Nephelium* sp.	Sapindaceae	8	4		0.0149	60.97	1.64
104	*Ormosia balansae* Drake.	Fabaceae	1			0.0154	67.05	0.22
105	*Ormosia fordiana* Oliv.	Fabaceae	1			0.0032	13.74	0.19
106	*Ormosia henryi* (Prain) Yakovlev	Fabaceae	4	2		0.0310	109.68	0.79
107	*Ormosia* sp.	Fabaceae	1		9	0.0082	23.10	0.20
108	*Palaquium annamense* Lecomte.	Sapotaceae	63	12	12	0.0240	144.62	10.76
109	*Palaquium* sp.	Sapotaceae	7	1	2	0.0390	231.14	1.70
110	*Paralbizia lucia*	Fabaceae	3	2	1	0.0064	22.42	0.48
111	*Parashorea* sp.	Dipterocarpaceae	3			0.0107	36.77	0.39
112	*Parinari annamensis* (Hance) J.E. Vidal	Chrysobalanaceae	3			0.0252	107.58	0.49
112	*Peltophorum tonkinensis* A.Chev	Caesalpiniaceae	1			0.0082	34.26	0.20
113	*Pometia pinnata* Forst. & Forst. f.	Sapindaceae	24	8	62	0.0226	105.50	4.64
114	*Pometia spp.*	Sapindaceae	14	4	68	0.0099	47.75	2.56
115	*Prunus arborea* (Blume) Kalkm.	Rosaceae	1			0.0092	38.54	0.21
116	*Pterospermum argenteum* Tardieu	Sterculiaceae	1			0.0032	6.25	0.19
117	*Pterospermum heterophyllum* Hance	Sterculiaceae	4			0.0132	48.56	0.62
118	*Quercus* sp.	Fagaceae		1				
119	*Rhus succeda* Wax.	Anacardiaceae	3			0.1097	882.08	1.22
120	*Sapium discolor* (Champ.) Muell.-Arg.	Euphorbiaceae	5			0.0310	202.33	1.05

No.	Tree Species	Plant Family	Counts			Basal Area (m²ha⁻¹)	Biomass (kg/tree)	Importance Value Index
			Mature	Sapling	Seedling	(m²ha⁻¹)	(kg/tree)	
121	*Sapium sebiferum* (L.) Roxb.	Euphorbiaceae	17		3	0.0797	501.27	5.27
122	*Saurauia tristyla* DC.	Actinidiaceae	17	1	1	0.0797	501.27	5.27
123	*Scaphium lychnophorum* (Hance.) Pierre.	Sterculiaceae	31	7	31	0.0803	613.28	9.83
124	*Schefflera glomerulata* H. L. Li	Araliaceae	2			0.0327	99.64	0.40
125	*Schefflera octophylla* (Lour.) Harms.	Araliaceae	34	10	18	0.0185	67.00	5.60
126	*Schefflera violea*	Araliaceae	16	4	7	0.0079	26.58	2.41
127	*Semecarpus anacardiopsis* Evrard & Tardieu	Anacardiaceae	1			0.0576		
128	*Sindora siamensis* Teijsm. *ex* Miq.	Caesalpiniaceae	2			0.0253	118.46	0.49
129	*Sindora tonkinensis* A. Chev. *ex* K. & S. Larsen.	Caesalpiniaceae	8	1	1	0.0236	84.85	1.57
130	*Sterculia lanceolata* Cav.	Sterculiaceae	2	3	1	0.0133	45.02	0.31
131	*Styrax annamensis* Guillaum.	Styracaceae	1	1		0.0032	12.25	0.19
132	*Styrax argentifolia*	Styracaceae	13		1	0.0324	209.98	2.56
133	*Styrax tonkinensis* (Pierre) Craib *ex* Hartwich	Styracaceae	2			0.0095	32.49	0.29
134	*Syzygium jambos* (L.) Alston.	Myrtaceae	19	7		0.0406	141.73	4.62
135	*Syzygium* sp.	Myrtaceae	24	13	5	0.0194	112.98	4.33
136	*Syzygium syzygioides* (Miq.) Merr. & Perry.	Myrtaceae	42	10	30	0.0110	47.04	6.19
137	*Syzygium zeylanicum*	Myrtaceae	1			0.0127	50.01	0.22
138	*Tarrietia javanica* Blume.	Sterculiaceae	10	5	15	0.0182	95.94	1.80
139	*Turpinia cochinchinensis*	Staphyleaceae	3			0.0171	69.31	0.56

Continued

No.	Tree Species	Plant Family	Counts			Basal Area (m²ha⁻¹)	Biomass (kg/tree)	Importance Value Index
			Mature	Sapling	Seedling			
140	*Vitex trifoliata*	Verbenaceae	17		6	0.0331	130.95	3.88
141	*Wrightia annamensis* Eberh. & Dub.	Apocynaceae	6			0.0221	109.42	1.18
142	*Xanthophyllum* sp.	Polygalaceae	9	1		0.0291	191.63	1.92
143	*Unknown A*		1			0.0127	61.41	0.22
144	*Unknown B*		1			0.0127	47.14	0.22
145	*Unknown C*		1			0.0241	119.31	0.24
146	*Unknown D*		1			0.0631	384.40	0.34
147	*Unknown E*		1			0.0087	32.44	0.21
148	*Unknown F*		1			0.0306	177.14	0.26
149	*Unknown G*		1			0.0199	99.13	0.23
150	*Unknown H*		1			0.0050	14.30	0.20
151	*Unknown I*		1			0.0072	17.07	0.20
152	*Unknown K*		2			0.0301	147.69	0.39
153	*Unknown L*			1				
154	*Unknown M*			1				

Index

Note: Page numbers followed by *f* indicate figures, *t* indicate tables, and *b* indicate boxes.

Printed in the United States
By Bookmasters